NONLINEAR FUNCTIONAL ANALYSIS

NONLINEAR FUNCTIONAL ANALYSIS

Rong Yuan
Beijing Normal University, China

高等教育出版社
HIGHER EDUCATION PRESS

World Scientific

Published by

World Scientific Publishing Co. Pte. Ltd.
5 Toh Tuck Link, Singapore 596224
USA office: 27 Warren Street, Suite 401-402, Hackensack, NJ 07601
UK office: 57 Shelton Street, Covent Garden, London WC2H 9HE

Library of Congress Cataloging-in-Publication Data
Names: Yuan, Rong, 1962– author
Title: Nonlinear functional analysis / author, Rong Yuan, (Beijing Normal University, China).
Description: Beijing, China : Higher Education Press Limited Company |
"Originally published in Chinese by Higher Education Press Limited Company in 2017"--
Copyright page. | Includes bibliographical references and index.
Identifiers: LCCN 2026008598 | ISBN 9789819825813 hardcover |
ISBN 9789819825820 ebook | ISBN 9789819825837 ebook other
Subjects: LCSH: Nonlinear functional analysis
Classification: LCC QA321.5 .Y836 2026
LC record available at https://lccn.loc.gov/2026008598

British Library Cataloguing-in-Publication Data
A catalogue record for this book is available from the British Library.

非线性泛函分析
Originally published in Chinese by Higher Education Press Limited Company

For any available supplementary material, please visit
https://www.worldscientific.com/worldscibooks/10.1142/14661#t=suppl

Desk Editors: Nambirajan Karuppiah/Angeline Husni

Typeset by Stallion Press
Email: enquiries@stallionpress.com

Preface

Many practical problems are nonlinear problems. The study of nonlinear problems has given rise to various research methods in nonlinear science today. This stimulates the emergence and development of nonlinear functional analysis theory. Nowadays, nonlinear functional analysis has become the foundation of the modern analysis for studying nonlinear problems.

There have been many books titled as nonlinear functional analysis, nonlinear analysis, degree theory, or variational methods, etc. And, many books on dynamical systems and partial differential equations also introduce some basic contents of nonlinear functional analysis. Usually, the textbooks of functional analysis for undergraduate students only include the main contents of linear functional analysis, while the current textbooks of functional analysis for graduate students also include some basic contents of nonlinear functional analysis. This indicates that it is essential necessary to teach the fundamental theory of nonlinear functional analysis in graduate courses.

Based on this consideration, the School of Mathematical Sciences at Beijing Normal University has made nonlinear functional analysis as a fundamental course for graduate students, and has also set up a course of the same name as an elective course for senior undergraduate students. This book is based on a Chinese version that was written by the author based on the lecture notes of teaching this fundamental course. The content of nonlinear functional analysis is vast and closely related to many problems. To give an introduction within a total of 48 class hours and 3 class hours per week, it is necessary to make a choice about the contents. We select differential calculus, implicit function theorem, degree theory, bifurcation theory, cone, and variational method as the teaching contents of the nonlinear functional analysis course. These contents are the most fundamentals and have become the basic knowledge required for studying nonlinear problems.

Books and textbooks under the name of nonlinear functional analysis can be partially understood from references. These books or textbooks provide a lot of materials and thinkings for the writing of this book. The study and research of nonlinear functional analysis need to be linked to the problem being studied. But, this book does not delve into this matter in depth, mainly due to time and space limitations. It is recommended that readers refer to other works.

During the writing process of this book, I received strong support from the School of Mathematical Sciences at Beijing Normal University, a lot of help from my family, as well as assistance from many colleagues and students. I received many valuable suggestions from the reviewing experts during the review process. Dr. Liangping Qi and Dr. Xiaofeng Zhang helped me to correct the manuscript. I would like to express my gratitude to them. Thanks are also due to Higher Education Press editor, Huaying Li, for her patience and professional assistance.

Finally, I acknowledge the funding support from the National Natural Science Foundation of China.

About the Author

Rong Yuan is a Professor and Doctoral supervisor at Beijing Normal University. He earned his bachelor's degree from the Department of Mathematics at Xuzhou Normal University, his master's degree from the Department of Mathematics at Nanjing University, and his doctoral degree from the Department of Mathematics at Peking University. He subsequently conducted postdoctoral research at the University of Cologne in Germany.

Professor Yuan's research primarily focuses on the qualitative theory of ordinary differential equations and dynamical systems. He has conducted in-depth studies in various areas, including qualitative analysis of almost periodic differential equations, qualitative analysis of non-autonomous dynamical systems, bifurcation problems in functional differential equations, nonlinear functional analysis and its applications in differential equation theory, as well as differential equation modeling and qualitative analysis in mathematical biology. To date, he has published more than 250 academic papers. Professor Yuan has long been engaged in teaching ordinary differential equations and nonlinear functional analysis, receiving widespread acclaim for his pedagogical excellence.

Since 1997, Professor Yuan has obtained numerous research grants, including the National Natural Science Foundation for Young Scholars, Key Projects, General Projects of the National Natural Science Foundation, and the Support Program for Outstanding Young

Teachers of the Ministry of Education. His notable awards include the Second Prize of the National Natural Science Award (Ministry of Education, 2003, independently), the Beijing Normal University Teaching Excellence Award (2013), the First Prize of Beijing Normal University Outstanding Graduate Teaching Achievement Award (2015), and the Beijing Municipal Higher Education Teaching Master Award (2020). His textbook *Ordinary Differential Equations* was recognized as a Beijing Higher Education Classic Textbook in 2013.

Contents

Chapter 1

Nonlinear Operators on Banach Spaces

We will first recall Banach spaces and linear operators. The contents of this book are mainly concerned with the nonlinear operators on infinite dimensional Banach spaces. In this chapter, we will discuss differential calculus in Banach spaces, implicit function theorem, bifurcation, cone and parting ordering in Banach spaces.

1.1 Banach Spaces and Linear Operators

In this section, we briefly survey some basic concepts and general results on the theory of linear functional analysis. A more detailed account can be found in the books [4, 5].

1.1.1 Banach Spaces and Hilbert Spaces

Definition 1.1.1. Let X be a nonempty set. If there is a real-valued function with two variables $\rho : X \times X \to \mathbb{R}$ such that

(1) $\rho(x, y) \geq 0$, and $\rho(x, y) = 0 \Longleftrightarrow x = y$;
(2) $\rho(x, y) = \rho(y, x)$;
(3) $\rho(x, z) \leq \rho(x, y) + \rho(y, z), \quad \forall x, y, z \in X$,

X is called a **metric space**, ρ is called a **distance**. A metric space X with a distance ρ is denoted by (X, ρ).

Definition 1.1.2. A sequence $\{x_n\}$ in a metric space (X, ρ) is said to be a **Cauchy sequence** if, for every $\epsilon > 0$, there is an integer $N = N(\epsilon)$ such that $\rho(x_n, x_m) < \epsilon$ as $m, n \geq N$.

If every Cauchy sequence in a space X converges to a limit in X, the space X is called a **complete space**.

Let K be the real number field or complex number field. Suppose that X is a vector space (or a linear space) over K and θ is the zero element of X.

Definition 1.1.3. A **norm** on a vector space X is a real-valued functional f on X satisfying the following conditions:

(1) $f(x) \geq 0$, $\forall x \in X$, and $f(x) = 0 \Longleftrightarrow x = \theta$ (positivity);
(2) $f(x+y) \leq f(x) + f(y)$, $\forall x, y \in X$ (triangle inequality);
(3) $f(\alpha x) = |\alpha| \, f(x)$, $\forall \alpha \in K, \forall x \in X$ (homogeneity).

We denote by $||\cdot||$ the norm, that is, let $||x|| = f(x)$ for all $x \in X$.

Definition 1.1.4. A **normed linear space** X is a vector space X provided with a norm.

A complete normed linear space is called a **Banach space**, for short, **B space**.

A distance in a normed linear space X is usually defined by

$$d(x, y) = ||x - y||, \quad x, y \in X.$$

Definition 1.1.5. Suppose that X is a vector space over a number filed K. A functional $\langle \cdot, \cdot \rangle$ defined on $X \times X \to K$ is called an **inner product** on X provided that for any $x, y, z \in X$ and $\alpha \in K$

(1) $\langle \alpha x, y \rangle = \alpha \langle x, y \rangle$;
(2) $\langle x + y, z \rangle = \langle x, z \rangle + \langle y, z \rangle$;
(3) $\langle x, y \rangle = \overline{\langle y, x \rangle}$ (conjugate symmetry);
(4) $\langle x, x \rangle \geq 0$, and $\langle x, x \rangle = 0 \Longleftrightarrow x = \theta$ (positivity).

Equipped with such a functional, X is called an **inner product space**. If $K = \mathbb{R}$, X is called a **real inner product space**; if $K = \mathbb{C}$, X is called a **complex inner product space**. The functional

$$||x|| := \sqrt{\langle x, x \rangle}, \quad x \in X$$

is a norm on X. $||\cdot||$ is called the induced norm by the inner product.

Definition 1.1.6. If X is an inner product space and complete (i.e. a Banach space) under the induced norm, we say that X is a **Hilbert space**.

1.1.2 Examples of Banach Spaces

Example 1.1. The space $C^k(\bar{\Omega})$.

Suppose that $\Omega \subset \mathbb{R}^n$ is a bounded domain and k is an integer. We define the vector space $C^k(\bar{\Omega})$ to be the space of all functions $f : \bar{\Omega} \to \mathbb{R}$ which are k-times continuously differentiable in $\bar{\Omega}$, where addition and scalar multiplication are defined in the natural manner. For every $u \in C^k(\bar{\Omega})$, the norm of u is defined as follows:

$$||u||_{C^k} = \max_{|\alpha| \leq k} \max_{x \in \bar{\Omega}} |\partial^\alpha u(x)|,$$

where $\alpha = (\alpha_1, \alpha_2, \ldots, \alpha_n)$ is a multi-index, $|\alpha| = \alpha_1 + \cdots + \alpha_n$, and

$$\partial^\alpha u(x) = \frac{\partial^{|\alpha|} u(x)}{\partial x_1^{\alpha_1} \cdots \partial x_n^{\alpha_n}}.$$

It is well known that it is an infinite dimensional Banach space.

The subspace $C_0^k(\Omega)$ consists of all those functions in $C^k(\bar{\Omega})$ that have compact support in Ω, where the support of u is defined by $\overline{\{x \in \bar{\Omega}; u(x) \neq 0\}}$.

Example 1.2. The Space $L^p(\Omega)$.

Suppose that $\Omega \subset \mathbb{R}^n$ is a Lebesgue measurable subset, $p \in [1, \infty)$. Let

$$L^p(\Omega) = \left\{ u;\ u \text{ is measurable function on } \Omega, \int_\Omega |u(x)|^p dx < \infty \right\},$$

where addition and scalar multiplication are defined in the natural manner. Two elements of $L^p(\Omega)$ are considered identical if they are equal almost everywhere. Thus, the elements of $L^p(\Omega)$ are actually not functions but equivalence classes of functions; two functions belong to the same element of $L^p(\Omega)$ if they are equal a.e. on Ω.

For every $u \in L^p(\Omega)$, the functional $||\cdot||$ defined by

$$||u|| = \left(\int_\Omega |u(x)|^p dx\right)^{\frac{1}{p}}$$

is a norm on $L^p(\Omega)$ provided $1 \le p < \infty$. It is not a norm if $0 < p < 1$.

Example 1.3. The space l^p.

Let

$$l^p = \left\{x = (x_1, x_2, \dots, x_k, \dots);\ x_k \in \mathbb{R}, \sum_{k=1}^{\infty} |x_k|^p < \infty\right\},$$

with the norm

$$||x|| = \left(\sum_{k=1}^{\infty} |x_k|^p\right)^{\frac{1}{p}}.$$

Example 1.4. The space $L^\infty(\Omega)$.

Suppose that $\Omega \subset \mathbb{R}^n$ is a measurable subset. A function u that is measurable on Ω is said to be **essentially bounded** on Ω if there is a constant K such that $|u(x)| \le K$ a.e. on Ω. The greatest lower bound of such constants K is called the essential supremum of $|u|$ on Ω, and is denoted by $\operatorname{ess\,sup}_{x\in\Omega} |u(x)|$. We denote by $L^\infty(\Omega)$ the vector space of all functions u that are essentially bounded on Ω, functions being identified if they are equal a.e. on Ω. The functional $||\cdot||_\infty$ defined by

$$||u||_\infty = \operatorname*{ess\,sup}_{x\in\Omega} |u(x)| = \inf_{\mu(E)=0, E\subset\Omega} \left(\sup_{x\in\Omega\setminus E} |u(x)|\right)$$

is a norm on $L^\infty(\Omega)$.

Example 1.5. The Sobolev space $W^{k,p}(\Omega)$.

Suppose that u, v are Lebesgue integrable and $\alpha = (\alpha_1, \dots, \alpha_n)$ is a multi-index. If for every $\phi \in C_0^{|\alpha|}(\Omega)$ there exists

$$\int_\Omega v\phi dx = (-1)^{|\alpha|} \int_\Omega u\partial^\alpha \phi dx,$$

we say that v is an α-order **weak derivative** of u, denoted by $v = \partial^\alpha u$. If there exists $\partial^\alpha u$ for any multi-index α, $|\alpha| \leq k$, we say that u is k-times weak differentiable. We denote by $W^k(\Omega)$ the set of all functions u that are k-times weak differentiable. Put

$$W^{k,p}(\Omega) = \{u \in W^k(\Omega);\ \partial^\alpha u \in L^p(\Omega), |\alpha| \leq k\},$$

and the norm

$$||u|| = \left(\int_\Omega \sum_{|\alpha| \leq k} |\partial^\alpha u|^p dx \right)^{\frac{1}{p}}.$$

$W_0^{k,p}(\Omega)$ that is the closure of $C_0^k(\Omega)$ in $W^{k,p}(\Omega)$ is a subspace of $W^{k,p}(\Omega)$.

1.1.3 Bounded Linear Operators

Suppose that X and Y are linear spaces, and $D \subset X$ is a subspace. $T : D \to Y$ is a mapping from D to Y, where D is called the **domain** of T and denoted by $D(T)$ sometimes. If

$$T(\alpha x + \beta y) = \alpha Tx + \beta Ty, \quad \forall x, y \in D, \forall \alpha,\ \beta \in K,$$

we say that T is a **linear operator**. $R(T) = \{Tx;\ \forall x \in D\}$ is called the **range** of T.

A linear operator is a certain kind of function whose domain is a linear space and whose range is contained in another linear space (possible the same as the first one). Thus, we may assume $D(T) = X$ without loss of generality.

Definition 1.1.7. Suppose that X and Y are Banach spaces. We say that the linear operator $T : X \to Y$ is **bounded**, if there is a constant M such that $||Tx||_Y \leq M||x||_X$, $\forall x \in X$.

Theorem 1.1. *Suppose that X and Y are Banach spaces, and $T : X \to Y$ is a linear operator. Then T is bounded if and only if T is continuous.*

Let $L(X, Y)$ be the set of bounded linear operators on X into Y. For a linear operator $T : X \to Y$, we define

$$||T|| = \sup_{||x||_X \leq 1} ||Tx||_Y.$$

$||T||$ is called the **norm** of operator T. For bounded linear operators, we have the following important theorems.

Theorem 1.2 (Banach). *Assume that X and Y are Banach spaces. If $T \in L(X, Y)$ is injective and surjective (bijective), then $T^{-1} \in L(Y, X)$.*

The general form of Banach theorem is the open mapping theorem as follows.

Theorem 1.3 (The Open Mapping Theorem). *Assume that X and Y are Banach spaces. If $T \in L(X, Y)$ is a surjective, then T is an open mapping.*

Theorem 1.4 (The Closed Graph Theorem). *Let X and Y be Banach spaces. Let $T : X \to Y$ be a closed linear operator and $D(T)$ be closed. Then T is continuous.*

Theorem 1.5 (Banach–Steinhaus, The Principe for Uniform Boundedness). *Suppose X and Y are Banach spaces. If*

$$W \subset L(X, Y), \quad \text{and} \sup_{T \in W} ||Tx|| < \infty \quad (\forall x \in X),$$

then there exists a constant M such that $||T|| \leq M$ $(\forall T \in W)$.

Let X be a real Banach space. A linear mapping from X into $\mathbb{R}$ is called a **linear functional** on X.

Theorem 1.6 (Hahn–Banach). *Suppose that X is a Banach space, X_0 is a subspace of X, and f_0 is a bounded linear functional defined on X_0. Then there exists a bounded linear functional defined on X such that*

(1) $f(x) = f_0(x) \quad (\forall x \in X_0)$ *(extension)*;
(2) $||f|| = ||f_0||_{X_0}$ *(norm preserving)*.

Let X be a real Banach space, X^* be the space of all continuous linear functionals on X, $f \in X^* \backslash \{0\}$, $r \in \mathbb{R}$. The subset

$$H = \{x \in X;\ f(x) = r\}$$

is called a **hyperplane** of X; the subsets

$$H^r(f) = \{x \in X;\ f(x) \geq r\}, \quad H_r(f) = \{x \in X;\ f(x) \leq r\}$$

are called **half spaces** determined by H, which lie on two sides of H.

Let A and B be two subsets of X. If there is a hyperplane H such that A lies one side of H and B lies on another side of H, then A and B are said to be **separable by hyperplane** H. In addition, if $A \cap H = \emptyset$ and $B \cap H = \emptyset$ hold too, A and B are **strictly separable by** H.

Suppose that A is a subset of X and H is a hyperplane of X. If A lies on one side of H and $H \cap A \neq \emptyset$, we say that H is a **supporting hyperplane of** A.

Theorem 1.7 (Mazur's Separation Theorem of Convex Sets). *Suppose that X is a real Banach space, $C \subset X$ is a convex subset, and interior $\overset{\circ}{C} \neq \emptyset$. Then the following two statements hold:*

(1) *Let $M = x_0 + X_0$, $x_0 \in X$, and X_0 be a subspace of X. If $M \cap \overset{\circ}{C} = \emptyset$, then there is a hyperplane H in X such that $M \subset H$ and $H \cap \overset{\circ}{C} = \emptyset$. In particular, for every $x_0 \in C \backslash \overset{\circ}{C}$, there exists a supporting hyperplane H of C such that $x_0 \in H$.*
(2) *Let $C_1 \neq \emptyset$ be another convex subset of X and $C_1 \cap \overset{\circ}{C} = \emptyset$. Then there is a hyperplane H separating C_1 and C.*

1.1.4 Dual Spaces

Theorem 1.8. *Let X^* be the space of all bounded linear functionals on X. Then X^* is a Banach space, where the norm is defined by*

$$||f|| = \sup_{||x|| \leq 1} |f(x)|, \quad \forall f \in X^*.$$

We say that X^* is the **dual space** of X. There are several convergences in a Banach space as follows.

Definition 1.1.8. Let X be a Banach space, $\{x_n\} \subset X$, and $x \in X$. If

$$\lim_{n\to\infty} f(x_n) = f(x), \quad \forall f \in X^*,$$

we say that $\{x_n\}$ **converges weakly** to x, denoted by $x_n \rightharpoonup x$.

Definition 1.1.9. Let $f_n,\ f \in X^*,\ n = 1, 2, \ldots$.

(1) If $||f_n - f|| \to 0$ $(n \to \infty)$, we say that f_n **converges uniformly** to f (or f_n converges to f in the norm).
(2) If $\lim_{n\to\infty} f_n(x) = f(x)$ for each $x \in X$, we say that f_n **converges $*$-weakly** to f.

Theorem 1.9 (F. Riesz). *Suppose that f is a continuous linear functional on a Hilbert space X. Then there exists a unique $y_f \in X$ such that*

$$f(x) = \langle x, y_f \rangle, \quad \forall x \in X.$$

Theorem 1.10 (Banach–Alaoglu). *If X is a separable Banach space, then every bounded sequence in X^* contains a $*$-weak convergent subsequence.*

Theorem 1.11 (Eberlein–Schmulyan). *Suppose that X is a Banach space.*

(1) *If X is reflexive, then each bounded sequence in X contains a weakly convergent subsequence.*
(2) *Conversely, if every bounded sequence in X contains a weakly convergent subsequence, then X is reflexive.*

The dual space of $\mathbb{R}^n$ is $\mathbb{R}^n$. The dual space of $L^p(\Omega)$ is $L^q(\Omega)$, where $\frac{1}{p} + \frac{1}{q} = 1$ $(p > 1)$, or $q = \infty$ $(p = 1)$.

Remark. In the space X^*, there are weak convergence and $*$-weak convergence. The weak convergence $x_n^* \rightharpoonup x^*$ means that $\langle x^{**}, x_n^* - x^* \rangle \to 0$ holds for every $x^{**} \in X^{**}$. The $*$-weak convergence means that $\langle x_n^* - x^*, x \rangle \to 0$ holds for every $x \in X$. Since $X \hookrightarrow X^{**}$ is continuous, weak convergence implies $*$-weak convergence.

1.1.5 Spectrum of Linear Operators

Let X be a complex Banach space, $T : D(T) \subset X \to X$ be a linear operator, and $R(T) \subset X$ be the range of T.

Definition 1.1.10. If there exists $x \in D(T)\backslash\{\theta\}$ such that

$$Tx = \lambda x,$$

$\lambda \in \mathbb{C}$ is called an **eigenvalue** of T and x is called the **eigenvector** of T corresponding to the eigenvalue λ.

If $T : D(T) \to R(T)$ is bijective, the inverse $T^{-1} : R(T) \to D(T)$ is linear.

Definition 1.1.11. Let $T : D(T) \subset X \to X$ be a linear operator. We say that T is a **regular operator**, if T^{-1} holds, $R(T) = X$, and T^{-1} is a bounded linear operator.

Definition 1.1.12. Let $\lambda \in \mathbb{C}$ be a complex number. If $(\lambda I - T)$ is a regular operator, we say that λ is the **regular value** of T. The set of all regular values of T is called the **resolvent set** of T, denoted by $\rho(T)$. The operator $R_\lambda(T) = (\lambda I - T)^{-1}$ is called the **resolvent operator** of T.

The **spectrum** of T is the set of all complex numbers not in $\rho(T)$, denoted by $\sigma(T)$.

If T is a closed linear operator, the spectrum $\sigma(T)$ is decomposed into disjoint sets $\sigma_p(T), \sigma_c(T), \sigma_r(T)$ with the following properties:

(1) $\sigma_p(T)$ is the set of complex numbers for which $(\lambda I - T)^{-1}$ does not exist. $\sigma_p(T)$ is called the **point spectrum** of T.
(2) $\sigma_c(T)$ is the set of complex numbers for which $(\lambda I - T)^{-1}$ exists and the range $R(\lambda I - A) \neq X$, but the closure $\overline{R(\lambda I - A)} = X$. $\sigma_c(T)$ is called the **continuous spectrum** of T.
(3) $\sigma_r(T)$ is the set of complex numbers for which $(\lambda I - T)^{-1}$ exists and $\overline{R(\lambda I - A)} \neq X$. $\sigma_r(T)$ is called the **residual spectrum** of T.

Thus, $\sigma(T) = \sigma_p(T) \cup \sigma_c(T) \cup \sigma_r(T)$.

Besides being divided into the point spectrum, the continuous spectrum, and residual spectrum, the spectrum of a linear operator can also be classified into discrete spectrum and essential spectrum that are defined as follows.

Definition 1.1.13. Let $T : D(T) \subset X \to X$ be a closed linear operator on the complex Banach space X. The **discrete spectrum** of T, denoted by $\sigma_d(T)$, is the set of $\lambda \in \sigma(T)$ such that the following two conditions hold:

(1) λ is an isolated point of $\sigma(T)$, that is, there is a neighborhood U of λ such that $U \cap \sigma(T) = \{\lambda\}$;
(2) $\text{Ker}(\lambda I - T)$ is finite dimensional, i.e., $\dim \ker(\lambda I - T) < \infty$.

Definition 1.1.14. Let $T : D(T) \subset X \to X$ be a closed linear operator on the complex Banach space X. The **essential spectrum** of T, denoted by $\sigma_{\text{ess}}(T)$, is the set of $\lambda \in \sigma(T)$ such that at least one of the following three conditions holds:

(1) λ is a limit point of $\sigma_p(T)$;
(2) $\text{Ker}(\lambda I - T)$ is infinite dimensional, i.e., $\dim \ker(\lambda I - T) = \infty$;
(3) $\lambda \in \sigma_c(T)$, i.e., the range $R(\lambda I - A) \neq X$, but the closure $\overline{R(\lambda I - A)} = X$.

1.1.6 Compact Operator and Riesz–Schauder Theory

Definition 1.1.15. Suppose that X and Y are Banach spaces, and $T : X \to Y$ is a linear operator. If, for every bounded subset B of X, $\overline{T(B)}$ is compact in Y, we say that T is a **compact operator**. A compact operator is also called **completely continuous**.

Theorem 1.12. *Suppose that a linear operator $T : X \to X$ is a compact operator. Then the following conclusions hold.*

(1) $0 \in \sigma(T)$, *except* $\dim X < \infty$;
(2) $\sigma(T)\backslash\{0\} = \sigma_p(T)\backslash\{0\}$;
(3) $\sigma_p(T)$ *has no accumulation point except possibly the point* $\lambda = 0$.

Theorem 1.13. *If $T : X \to X$ is a compact linear operator, there exists a nonnegative integer p such that*

$$X = N((I - T)^p) \oplus R((I - T)^p),$$

and the operator $T_1 := T\big|_{R((I-T)^p)}$ *has a bounded inverse, where*

$$N((I - T)^p) := \{x;\ (I - T)^p x = 0\},$$

and $R((I - T)^p)$ *means the range.*

1.1.7 Poincaré Inequality and Sobolev Imbedding Theorem

Theorem 1.14 (Poincaré Inequality). *If $\Omega \subset \mathbb{R}^n$ is a bounded domain and $1 \leq p < \infty$, there is a constant $C = C(p, \Omega)$ such that*

$$\int_\Omega |u|^2 dx \leq C \int_\Omega |\nabla u|^2 dx, \quad \forall u \in W_0^{1,p}(\Omega).$$

Theorem 1.15 (Sobolev Imbedding Theorem). *Suppose that $\Omega \subset \mathbb{R}^n$ is a bounded domain with uniform C^m boundary, $1 \leq q < \infty$, $m \geq 0$ is an integer. Then the imbedding mapping*

$$W^{m,q}(\Omega) \hookrightarrow L^r(\Omega), \quad \frac{1}{r} \geq \frac{1}{q} - \frac{m}{n}\ (mq < n),$$

and $\forall j \in \mathbb{N}$,

$$W^{m+j,q}(\Omega) \hookrightarrow C^{j,\lambda}(\bar{\Omega}), \quad 0 < \lambda \leq m - \frac{n}{q}\ (mq > n)$$

are continuous.

In the case $m = 1$, letting $q^* = \frac{nq}{n-q}$, we obtain

$$W^{1,q}(\Omega) \hookrightarrow L^r(\Omega), \quad r \leq q^*\ (n > q),$$

and

$$W^{1,q}(\Omega) \hookrightarrow C(\bar{\Omega}) \quad (q > n)$$

are continuous.

Theorem 1.16 (Rellich–Kondrachov Compact Imbedding Theorem). *Suppose that $\Omega \subset \mathbb{R}^n$ is a bounded domain with uniform C^m boundary, $1 \leq q < \infty$, and $m \geq 0$ is an integer. Then the imbedding mapping*

$$W^{m,q}(\Omega) \hookrightarrow L^r(\Omega), \quad 1 \leq r < \frac{nq}{n - mq} \left(m < \frac{n}{q}\right),$$

and

$$W^{m,q}(\Omega) \hookrightarrow C(\bar{\Omega}) \left(m > \frac{n}{q}\right)$$

are compact.

1.2 The Calculus of Abstract Functions

1.2.1 The Integral of Abstract Functions

Let $f: I \to V$, where $I \subset \mathbb{R}^n$ is an interval or a measurable set, and V is a Banach space. Such a function is called **abstract function**.

A function $f: I \to V$ is called a **simple function**, if there exist countable disjoint measurable sets $E_j \subset I$, $j = 1, 2, \ldots$, Lebesgue measures $\mu(E_j) < \infty$, and vectors $v_j \in V$, $j = 1, 2, \ldots$, such that $f(t) = v_j$ as $t \in E_j$, while $f(t) = 0$ as $t \in I \backslash (\bigcup_j E_j)$.

Let $I \subset \mathbb{R}^n$ be a measurable subset and $\mu(I) < \infty$. We say that a simple function $f: I \to V$ is **Bochner integrable**, if $||f(t)||: I \to \mathbb{R}$ is Lebesgue measurable. At this time, Bochner integral is defined by

$$\int_I f(t)dt = \sum_{j=1}^{\infty} v_j \mu(E_j).$$

We say that an abstract function $f: I \to V$ is **Bochner integrable**, if there exists a sequence of Bochner integrable simple functions $\{f_n\}: I \to V$, such that

(i) $\lim_{n\to\infty} f_n(t) = f(t)$ is strongly convergent, for *a.e.* $t \in I$;
(ii) $\lim_{n\to\infty} \int_I ||f_n(t) - f(t)||dt = 0$, here the integral is Lebesgue integral.

The Bochner integral of f on I is defined by

$$\int_I f(t)dt = \lim_{n\to\infty} \int_I f_n(t)dt,$$

which is the limit of Bochner integral of f_n.

Theorem 1.17. *If $f: I \to V$ is Bochner integrable, then the following properties hold*

(1) $|| \int_I f(t)dt|| \leq \int_I ||f(t)||dt.$
(2) *Let $A: D(A) \subset V \to Y$ be a bounded linear operator. If $f: I \to V$ is Bochner integrable and $Af: I \to Y$ is Bochner integrable, then $\int_I f(t)dt \in D(A)$ and*

$$A \int_I f(t)dt = \int_I Af(t)dt;$$

(3) **(Dominated Convergence Theorem)** *Suppose that* $\{f_n\} : I \to V$ *are Bochner integrable,* $f : I \to V$*, and*

(a) $\lim_{n\to\infty} f_n(t) = f(t)$ *is strongly convergent, for a.e.* $t \in I$;
(b) *there is a Lebesgue integrable function* $p(t)$ *on* I *such that*

$$||f_n(t)|| \leq p(t), \quad a.e.\, t \in I, \forall n.$$

Then $f : I \to V$ *is Bochner integrable and*

$$\lim_{n\to\infty} \int_I f_n(t)dt = \int_I f(t)dt.$$

(4) **(Fubini's Theorem)** *Let* $I \subset \mathbb{R}$ *and* $J \subset \mathbb{R}$ *be measurable subsets. If* $f(t,s) : I \times J \to V$ *is a Bochner integrable function, then*

$$\int_{I\times J} f(t,s)dtds = \int_I \left(\int_J f(t,s)ds \right) dt = \int_J \left(\int_I f(t,s)dt \right) ds.$$

1.2.2 The Differentiation of Abstract Functions

Let $f : I \to V$, where $I \subset \mathbb{R}$ is an interval and V is a Banach space. We say that the function f is **strongly differentiable** at the point $t_0 \in I$, if there is a vector $v_0 \in V$ such that

$$f(t_0 + h) - f(t_0) = hv_0 + E(v_0, h), \quad t_0 + h \in I,$$

and

$$h^{-1}||E(v_0, h)|| \to 0, \quad |h| \to 0.$$

Put $v_0 = f'(t_0) = \partial_t f\big|_{t=t_0} = \partial_t f(t_0)$. We say that $\partial_t f$ is a **(strong) derivative** of f.

Let $f : I = [a,b] \to V$. The function f on I is said to be **absolutely continuous** if, for every $\epsilon > 0$, there exists a $\delta(\epsilon) > 0$ such that

$$\sum_{n=1}^{N} ||f(t_n) - f(s_n)||_V < \epsilon,$$

whenever $\{(s_1, t_1), \ldots, (s_N, t_N)\}$ is a finite collection of disjoint intervals of total length $\sum_{n=1}^{N} |t_n - s_n|$ at most δ.

Theorem 1.18 (Newton–Leibniz's Formula). *Let V be a Banach space. Suppose that $f : [a,b] \to V$ is an absolutely continuous function and its strong derivative f' is Bochner integrable on $[a,b]$, i.e., $f' \in L^1([a,b];V)$. Then we have*

$$f(t) = f(a) + \int_a^t f'(s)ds, \quad t \in [a,b].$$

1.3 Fréchet Differentiation

In this section, we will consider nonlinear operators defined on a Banach space. Let X and Y be two Banach spaces, $||\cdot||$ denote its norm for no confusing, $f : D \to Y$ be a mapping, and $D \subset X$ be the domain of f.

Definition 1.3.1. Let $x_0 \in D$. We say that f is **continuous** at x_0, if, for every $\epsilon > 0$, there exists $\delta = \delta(x_0, \epsilon) > 0$ such that $||f(x) - f(x_0)|| < \epsilon$ whenever $x \in D$ and $||x - x_0|| < \delta$. If δ is independent of $x_0 \in D$ and depends on ϵ only, we say that f is **uniformly continuous** on D.

Clearly, f is continuous at $x_0 \in D \iff \forall \{x_n\} \subset D$, we have $f(x_n) \to f(x_0)(n \to \infty)$ whenever $x_n \to x_0$.

Definition 1.3.2. Let $D \subset X$. We say that a mapping $F : D \to Y$ is a **compact mapping**, if $F(S)$ is relatively compact in Y, i.e. $\overline{F(S)}$ is a compact subset, for every bounded subset S of D. In addition, if F is continuous too, we say that F is a **compact continuous** mapping or **completely continuous** mapping.

Tietze's extension theorem is about that a continuous mapping on a closed subset can be extended to continuously those on all spaces. A general form is the following theorem.

Theorem 1.19 (J. Dugundji). *Let A be a closed subset in a metric space X and Y be a normed linear space. If $f : A \to Y$ is continuous, then there is a continuous extension $\tilde{f} : X \to \overline{\text{conv}} f(A)$ of f, where $\overline{\text{conv}} f(A)$ is the* **closed convex hull** *of $f(A)$, i.e., it is defined as the intersection of all closed convex subsets which contain D.*

If f is completely continuous, then f has a completely continuous extension $\tilde{f}$.

Definition 1.3.3. Let the mapping $f : U \subset X \to Y$, $x_0 \in U$. We say that f is **Fréchet differentiable** (for short, ***F*-differentiable** or **differentiable**) at x_0, if there is a bounded linear operator $A \in L(X,Y)$ such that whenever $h \in X$, $x_0 + h \in U$, we have

$$f(x_0 + h) - f(x_0) = Ah + w(x_0, h),$$

where $w(x_0, h) = o(||h||)$, i.e.,

$$\lim_{||h|| \to 0} \frac{||w(x_0, h)||}{||h||} = 0.$$

At this time, we say that A is the **Fréchet (or *F*-) derivative** of f at x_0, and is denoted by $df(x_0)$, $\nabla f(x_0)$ or $f'(x_0)$.

Clearly, if f is F-differentiable at x_0, the mapping $A \in L(X,Y)$ is unique and f is continuous at x_0. If f and g are F-differentiable at x_0, the linear combinations of f and g are F-differentiable at x_0. If $A \in L(X,Y)$ and $f(x) = Ax$, then $f'(x) = A$, $\forall x \in X$.

Theorem 1.20 (Chain Rule). *Let X, Y, Z be Banach spaces, and $U \subset X, V \subset Y$ be open subsets. If $f : U \to V$ is Fréchet differentiable at x_0 and $g : V \to Z$ is Fréchet differentiable at $y_0 = f(x_0)$, then $g \circ f : U \to Z$ is Fréchet differentiable at x_0, and*

$$d(g \circ f)(x_0) = dg(y_0) \circ df(x_0).$$

Proof. From the definition, for $h \in X$ and $k \in Y$, we have

$$f(x_0 + h) - f(x_0) - df(x_0)h = w(x_0, h) = o(||h||),$$
$$g(y_0 + k) - g(y_0) - dg(y_0)k = \theta(y_0, k) = o(||k||).$$

Let $k = df(x_0)h + w(x_0, h)$. Since $df(x_0)$ is a bounded linear operator, we obtain $||k|| \to 0$ as $||h|| \to 0$. Thus,

$$\begin{aligned}
&g \circ f(x_0 + h) - g \circ f(x_0) \\
&\quad = g(f(x_0) + df(x_0)h + w(x_0, h)) - g \circ f(x_0) \\
&\quad = dg(y_0)(df(x_0)h + w(x_0, h)) + \theta(y_0, k) \\
&\quad = dg(y_0) \circ df(x_0)h + dg(y_0)w(x_0, h) + \theta(y_0, k).
\end{aligned}$$

Since $dg(y_0)$ is a bounded linear operator, we have $||dg(y_0) \cdot w(x_0, h)|| = o(||h||)$. From

$$\begin{aligned}\frac{||\theta(y_0,k)||}{||h||} &= \frac{||\theta(y_0,k)||}{||k||} \cdot \frac{||df(x_0)h + w(x_0,h)||}{||h||} \\ &\le \frac{||\theta(y_0,k)||}{||k||} \cdot \left(||df(x_0)|| + \frac{||w(x_0,h)||}{||h||}\right) \\ &\le o(1)(||df(x_0)|| + o(1)),\end{aligned}$$

using the finiteness of $||df(x_0)||$, we have

$$||\theta(y_0,k)|| \le o(||h||).$$

So,

$$g \circ f(x_0 + h) - g \circ f(x_0) - dg(y_0) \circ df(x_0)h = o(||h||).$$

Therefore, $g \circ f$ is Fréchet differentiable at x_0 and

$$d(g \circ f)(x_0) = dg(y_0) \circ df(x_0).$$

This completes the proof.

We now examine the case where one of the two spaces X and Y is a product, equipped with any norm that induces the product.

Theorem 1.21. *Suppose that the space Y is a product $Y = Y_1 \times Y_2 \times \cdots \times Y_m$ of normed vector spaces Y_i, $1 \le i \le m$, a mapping $f : U \subset X \to Y$ is defined by m component mappings $f_i : U \subset X \to Y_i$, $1 \le i \le m$, that is, $f(x) = (f_1(x), \ldots, f_m(x)), x \in U$. Then the mapping f is Fréchet differentiable at a point $x_0 \in U$ if and only if each mapping $f_i, 1 \le i \le m$, is Fréchet differentiable at the same point x_0.*

At this time, the Fréchet derivative $f'(x_0) \in L(X,Y)$ can be identified with the element $(f_1'(x_0), f_2'(x_0), \ldots, f_m'(x_0))$ of the product space $L(X,Y_1) \times L(X,Y_2) \times \cdots \times L(X,Y_m)$.

Proof. To fix ideas, assume that the norm of $y = (y_1, \ldots, y_m) \in Y$ is defined by $||y|| = \max_{1 \le i \le m} ||y_i||_{Y_i}$.

If $f = (f_1, \ldots, f_m) : U \subset X \to Y$ is Fréchet differentiable at $x_0 \in U$, the relation $f(x_0 + h) - f(x_0) = f'(x_0)h + w(x_0, h)$ is equivalent to the m relations

$$f_i(x_0 + h) - f_i(x_0) = A_i h + w_i(x_0, h), \quad 1 \leq i \leq m,$$

where $A_i \in L(X, Y_i)$ is the ith component of $f'(x_0) \in L(X, Y)$. Since $||w_i(x_0, h)||_{Y_i} \leq ||w(x_0, h)||$, it follows that each mapping $f_i : U \subset X \to Y_i$ is Fréchet differentiable at x_0 with $f_i'(x_0) = A_i$, $1 \leq i \leq m$.

If each component mapping f_i, $1 \leq i \leq m$, is Fréchet differentiable at $x_0 \in U$, then

$$\begin{aligned}
&f(x_0 + h) - f(x_0) \\
&\quad = (f_1(x_0 + h) - f_1(x_0), \ldots, f_m(x_0 + h) - f_m(x_0)) \\
&\quad = (f_1'(x_0)h + w_1(x_0, h), \ldots, f_m'(x_0)h + w_m(x_0, h)) \\
&\quad = (f_1'(x_0)h, \ldots, f_m'(x_0)h) + (w_1(x_0, h), \ldots, w_m(x_0, h)).
\end{aligned}$$

Since

$$\max_{1 \leq i \leq m} ||f_i'(x_0)h||_{Y_i} \leq \left(\max_{1 \leq i \leq m} ||f_i'(x_0)|| \right) ||h||, \quad \forall h \in X,$$

we have that the linear operator $h \in X \to (f_1'(x_0)h, \ldots, f_m'(x_0)h) \in Y$ is continuous. From

$$\max_{1 \leq i \leq m} \frac{||w_i(x_0, h)||}{||h||} \to 0, \quad \text{as}\, h \to 0,$$

we know that f is Fréchet differentiable at x_0 and

$$f'(x_0) = (f_1'(x_0), \ldots, f_m'(x_0)).$$

This completes the proof.

Consider next the case where the space X is a product $X = X_1 \times X_2 \times \cdots \times X_n$ of normed victor spaces X_j, $1 \leq j \leq n$. Given a point $x_0 = (x_{10}, \ldots, x_{n0}) \in U$, there exists for each index j an open subset U_j of the space X_j containing the point x_{j0} such that the open set $U_1 \times \cdots \times U_n$ is contained in U. If, for some $1 \leq j \leq n$, the partial mapping

$$f(x_{10}, \ldots, x_{(j-1)\,0}, \cdot, x_{(j+1)\,0}, \ldots, x_{n0}) : U_j \subset X_j \to Y$$

is Fréchet differentiable at the point $x_{j0} \in U_j$, its Fréchet derivative is denoted as

$$\partial_j f(x_0) \in L(X_j, Y)$$

and is called the **jth partial Fréchet derivative** of the mapping f at the point x_0. If x_j denotes a generic point in the spaces X_j, such a partial Fréchet derivative is also denoted

$$\frac{\partial f}{\partial x_j}(x_0) := \partial_j f(x_0).$$

Theorem 1.22. *If a mapping $f : U \subset X = X_1 \times \cdots \times X_n$ is Fréchet differentiable at a point $x_0 \in U$, then the n partial Fréchet derivatives $\partial_j f(x_0)$, $1 \leq j \leq n$, exist and*

$$f'(x_0)h = \sum_{j=1}^{n} \partial_j(x_0)h_j, \quad \forall h = (h_1, \dots, h_n) \in X_1 \times \cdots \times X_n.$$

Proof. To fix ideas, assume that X is equipped with the norm $||x|| = \max\{||x_1||_{X_1}, \dots, ||x_n||_{X_n}\}$ for $x = (x_1, \dots, x_n) \in X$.

Since the Fréchet derivative $f'(x_0) \in L(X, Y)$ exists, we define

$$\begin{aligned} A_1 h_1 &:= f'(x_0)(h_1, 0, \dots, 0), \quad \forall h_1 \in X_1, \\ A_2 h_2 &:= f'(x_0)(0, h_2, 0, \dots, 0), \quad \forall h_2 \in X_2, \\ &\vdots \\ A_n h_n &:= f'(x_0)(0, \dots, 0, h_n), \quad \forall h_n \in X_n. \end{aligned}$$

It is easy to show that $A_1 \in L(X_1, Y), \dots, A_n \in L(X_n, Y)$, *i.e.*, they are continuous linear operators and

$$f'(x_0)h = A_1 h_1 + \cdots + A_n h_n, \quad \forall h = (h_1, \dots, h_n) \in X_1 \times \cdots \times X_n.$$

Since

$$\begin{aligned} f(x_{10} + h_1, x_{20}, \dots, x_{n0}) &= f(x_{10}, x_{20}, \dots, x_{n0}) + f'(x_0)(h_1, 0, \dots, 0) \\ &\quad + w(x_0, (h_1, 0, \dots, 0)) \end{aligned}$$

and

$$\frac{||w(x_0,(h_1,0,\ldots,0))||}{||h_1||_{X_1}} = \frac{||w(x_0,(h_1,0,\ldots,0))||}{||(h_1,0,\ldots,0)||_X} \to 0,$$
$$\text{as } ||h_1||_{X_1} \to 0,$$

it follows that $A_1 = \partial_1 f(x_0)$. Similar arguments show that $A_2 = \partial_2 f(x_0), \ldots, A_n = \partial_n f(x_0)$. This completes the proof.

Finally, suppose that

$$X = X_1 \times X_2 \times \cdots \times X_n, \quad Y = Y_1 \times Y_2 \times \cdots \times Y_m,$$

so that in this case a function $f : U \subset X \to Y$ is determined by means of m functions $f_i : U \subset X \to Y_i$ of n variables. Then the relation

$$k = f'(a)h, \quad h = (h_1, \ldots, h_n) \in X, \quad k = (k_1, \ldots, k_m) \in Y$$

is equivalent to the relations

$$k_i = \sum_{j=1}^{n} \partial_j f_i(x_0) h_j, \quad 1 \le i \le m.$$

1.4 Gâteaux Differentiation

Usually, we compute F-derivative by means of Gâteaux derivative or G-derivative, because it is not easy to write down the F-derivative for a given map directly.

Definition 1.4.1. We say that a mapping $f : U \to Y$ is **Gâteaux differentiable** (G-differentiable or weak differentiable for short) in direction $h \in X$ at $x_0 \in U$, if the limit

$$\lim_{t\to 0} \frac{f(x_0+th) - f(x_0)}{t} =: Df(x_0; h)$$

exists. $Df(x_0; h)$ is called the **Gâteaux differentiation** (G-differentiation or weak differentiation for short) in direction h at x_0. If f is Gâteaux differentiable in any direction at x_0, we say that f is **Gâteaux differentiable** (G-differentiable or weak differentiable for short) at x_0.

Definition 1.4.2. If $f : U \to Y$ is G-differentiable at $x_0 \in U$, and $Df(x_0; h)$ is bounded linear about h, i.e.,

$$Df(x_0; h) = Df(x_0)h, \quad Df(x_0) \in L(X, Y),$$

$Df(x_0)$ is called the **Gâteaux derivative** (G-derivative for short) of f at x_0.

From the definition, it is easily known that $Df(x_0; h)$ is homogeneous with respect to h, *i.e.*,

$$Df(x_0; sh) = sDf(x_0; h), \quad \forall s \in \mathbb{R},$$

and F-differentiable $\Longrightarrow$ G-derivative $\Longrightarrow$ G-differentiable.

Generally speaking, $Df(x_0; h)$ may not be linear with respect to h. For example, we consider

$$f(x) = \begin{cases} \frac{x_1^2 x_2}{x_1^2 + x_2^2}, & x = (x_1, x_2) \neq (0,0), \\ 0, & x = (x_1, x_2) = (0,0). \end{cases}$$

Its G-differentiation in direction $h = (h_1, h_2)$ at $x_0 = (0,0)$ is

$$Df(x_0; h) = \frac{h_1^2 h_2}{h_1^2 + h_2^2},$$

which is homogeneous on h, but not linear.

The existence of G-derivative does not imply F-differentiable. For example,

$$f(x) = \begin{cases} x_1 + x_2 + \frac{x_1^3 x_2}{x_1^4 + x_2^2}, & x = (x_1, x_2) \neq (0,0), \\ 0, & x = (x_1, x_2) = (0,0) \end{cases}$$

is G-differentiable at $(0,0)$ and has G-derivative, but is not F-differentiable at $(0,0)$.

G-differentiation is a generalization of the derivative in directions in the differential calculus of several variables.

Example 1.6. Let f be a real function on $\mathbb{R}^n$, and e_j be the jth standard base vector in $\mathbb{R}^n$. Then the G-differentiation in direction e_j at x is $Df(x; e_j) = \frac{\partial f(x)}{\partial x_j}$.

When Y is a product, a mapping is thus differentiable at a point if and only if all its component mappings are differentiable at the same point. By contrast, when X is a product, a mapping may no longer be differentiable at a point if all the partial mappings are differentiable at the same point.

If f has G-derivative at every point in U, Df is usually a nonlinear mapping from U into $L(X, Y)$. If $Df(x_0)$ exists, we have

$$f(x_0 + th) - f(x_0) = Df(x_0)th + w(x_0, h, t),$$

where

$$\lim_{t \to 0} \frac{w(x_0, h, t)}{t} = 0,$$

for every fixed x_0, h. Generally speaking, this limit is not uniform on $||h|| = 1$. However, we have the following result.

Theorem 1.23. *If the G-derivative of f at x_0 exists and the limit*

$$\lim_{t \to 0} \frac{1}{t}[f(x_0 + th) - f(x_0)] = Df(x_0)h$$

exists uniformly on $||h|| = 1$, then f is F-differentiable at x_0.

Proof. For every $\epsilon > 0$, it follows from the assumption that there exists a $\delta > 0$ independently of $||h|| = 1$, such that whenever $|t| < \delta$, we have

$$\left\| \frac{1}{t}[f(x_0 + th) - f(x_0)] - Df(x_0)h \right\| < \epsilon.$$

For every h_1, let $h_1 = th$, $t = ||h_1||$, $||h|| = 1$. If $||h_1|| < \delta$, we have

$$||f(x_0 + h_1) - f(x_0) - Df(x_0)h_1|| \leq \epsilon ||h_1||.$$

From the definition, we know that $df(x_0)$ exists and equals to $Df(x_0)$. This completes the proof.

Corollary 1.1. *If Gâteaux derivative mapping Df is continuous at x_0, then f is Fréchet differentiable at x_0.*

Proof. Because Df is continuous at x_0, for every $\epsilon > 0$, there is a $\delta > 0$ such that whenever $x \in U$ and $||x - x_0|| < \delta$, we have

$$\sup_{||h||=1} ||Df(x)h - Df(x_0)h|| < \epsilon.$$

Using the Mean Value Formula below, if $|t| < \delta$, for any $||h|| = 1$, we have

$$\begin{aligned} &||f(x_0 + th) - f(x_0) - Df(x_0)th|| \\ &= \left\| \int_0^1 Df(x_0 + sth)thds - \int_0^1 Df(x_0)thds \right\| \\ &\leq |t| \int_0^1 ||Df(x_0 + sth)h - Df(x_0)h||ds \\ &< |t|\epsilon. \end{aligned}$$

From Theorem 1.23, f is Fréchet differentiable at x_0. This is the end.

The computation of G-derivative may be reduced to the derivative of single variable mapping and can be obtained by using Taylor formula. Corollary 1.1 provides an approach to compute F-derivative through G-derivative. The following theorem is about G-differentiation $Df(x_0; h)$, so that the results hold for G-derivative and F-derivative.

Theorem 1.24. *The following statements hold for Gâteaux differentiation (G-derivative and F-derivative):*

(1) *If f is G-differentiable in direction h at $x_0 \in U$, then for any $y^* \in Y^*$, the function $\phi(t) = y^*(f(x_0 + th))$ is differentiable at $t = 0$, and $\phi'(0) = y^*(Df(x_0; h))$. If f is G-differentiable in direction h at every point in the line segment $L = \{x_0 + th;\ t \in [0, 1]\}$, then $\phi(t)$ is differentiable on $[0, 1]$, and*

$$\phi'(t) = y^*(Df(x_0 + th; h)).$$

(2) *If f is G-differentiable on the line segment L, there exists a $\tau \in [0, 1]$ such that*

$$||f(x_0 + h) - f(x_0)|| \leq ||Df(x_0 + \tau h; h)||.$$

(3) **(The Mean Value Theorem)** *If Y is a Banach space and $\psi(t) = Df(x_0 + th; h)$ is continuous on $[0, 1]$, then we have*

$$f(x_0 + h) - f(x_0) = \int_0^1 \psi(t)dt,$$

where $\int_0^1 \psi(t)dt$ is Riemann integral of the continuous abstract function $\psi(t)$.

Proof. Clearly,

$$\frac{\phi(t) - \phi(0)}{t} = y^* \left(\frac{f(x_0 + th) - f(x_0)}{t} \right).$$

(1) The conclusion follows directly from the definition.
(2) From the Hahn–Banach Theorem, it follows that there is a $y^* \in Y^*$, $||y^*|| = 1$ such that

$$y^*(f(x_0 + h) - f(x_0)) = ||f(x_0 + h) - f(x_0)||.$$

By (1) and the Mean Value Theorem, we have

$$\begin{aligned} ||f(x_0 + h) - f(x_0)|| &= \phi(1) - \phi(0) = \phi'(\tau) \\ &= y^*(Df(x_0 + \tau h; h)) \\ &\leq ||Df(x_0 + \tau h; h)||. \end{aligned}$$

(3) For every $y^* \in Y^*$, from the assumptions we know that $\phi(t) = y^*(f(x_0 + th))$ is differentiable on $[0, 1]$ and $\phi'(t) = y^*(\psi(t))$ is continuous on $[0, 1]$. From the Fundamental Theorem of Calculus, we obtain

$$\begin{aligned} y^*(f(x_0 + h) - f(x_0)) = \phi(1) - \phi(0) &= \int_0^1 \phi'(t)dt \\ &= \int_0^1 y^*(Df(x_0 + th; h))dt \\ &= y^* \left(\int_0^1 Df(x_0 + th; h)dt \right). \end{aligned}$$

By the arbitrariness of y^*, we have

$$f(x_0 + h) - f(x_0) = \int_0^1 Df(x_0 + th; h)dt.$$

This is the end.

Example 1.7. Let $f : U \subset \mathbb{R}^n \to \mathbb{R}^m$ can be written as $y = f(x)$, where $x = (x_1, \ldots, x_n)$, $y = (y_1, \ldots, y_m)$, $y_i = f_i(x_1, \ldots, x_n)$, $1 \leq i \leq m$. If f_i, $1 \leq i \leq m$, have continuous partial derivative in a neighborhood of a point $x^{(0)} = (x_1^{(0)}, \ldots, x_n^{(0)})$, then f is F-differentiable at $x^{(0)}$, and $f'(x^{(0)}) = \left(\frac{\partial f_i}{\partial x_j}(x^{(0)})\right)_{m\times n}$.

The mentioned above result is well known in differential calculus of several variables. This result can be extended to Banach spaces as follows.

Theorem 1.25. *Let $X_j, 1 \leq j \leq n$, and Y be normed vector spaces, U be an open subset of the product $X_1 \times \cdots \times X_n$, and $f : U \to Y$ be a mapping. Then f has continuous F-derivative if and only if all partial mappings defined by f have continuous partial F-derivatives.*

1.5 Examples

Let $G \subset \mathbb{R}^n$ be a Lebesgue measurable subset, $0 < \operatorname{mes} G < \infty$, and $f(x, u) : G \times \mathbb{R} \to \mathbb{R}$ be a function. If $u(x)$ is a function defined on G, then $f(x, u(x))$ is a function defined on G too. We say that the mapping

$$\mathbf{f} : u(x) \to f(x, u(x))$$

is the **Nemytskii operator**.

1.5.1 Continuity of Nemytskii Operators

Definition 1.5.1. We say that $f : G \times \mathbb{R} \to \mathbb{R}$ is a **Caratheodory function**, if

(i) $\forall$ *a.e.* $x \in G$, $f(x, u)$ is a continuous function in u;
(ii) $\forall u \in \mathbb{R}$, $f(x, u)$ is a measurable function for x.

Lemma 1.1. *Let $f : G\times\mathbb{R}\to\mathbb{R}$ be a Caratheodory function. If $u(x)$ is a Lebesgue measurable function on G, then so is $f(x,u(x))$.*

Proof. Since $u(x)$ is a Lebesgue measurable function on G, there is a sequence of simple functions $\{u_n(x)\}$ such that $u(x) = \lim_{n\to\infty} u_n(x)$ for almost everywhere $x\in G$. From (ii), we know that $f(x,u_n(x))$ is Lebesgue measurable for x. By (i), $f(x,u_n(x))\to f(x,u(x))$ for almost everywhere $x\in G$. Thus, $f(x,u(x))$ is Lebesgue measurable for x. This is the end.

Lemma 1.2. *Let $f : G\times\mathbb{R}\to\mathbb{R}$ be a Caratheodory function. If $u_n(x)$ converges to $u(x)$ in measure, then $f(x,u_n(x))$ converges to $f(x,u(x))$ in measure too.*

Proof. From the Riesz Theorem, we know that a sequence of measurable functions $\{g_n\}$ converges to g in measure on G, if and only if, for every subsequence $\{g_{n_k}\}$ of $\{g_n\}$, there exists a subsequence $\{g_{n_{k_\nu}}\}$ converging to g almost everywhere on G.

By the fact mentioned above, for every subsequence $\{f(x,u_{n_k}(x))\}$, there exists a subsequence $\{u_{n_{k_\nu}}(x)\}$ converging to $u(x)$ almost everywhere, because $u_n(x)$ converges to $u(x)$ in measure. From (i), we know that $\{f(x,u_{n_{k_\nu}}(x))\}$ converges to $f(x,u(x))$ almost everywhere. This is the end.

Theorem 1.26. *Let $f : G\times\mathbb{R}\to\mathbb{R}$ be a Caratheodory function and satisfy*

$$|f(x,u)|\le a|u|^{p_1/p_2}+b(x),\quad x\in G,\ u\in\mathbb{R}, \tag{1.1}$$

where $a>0$, $p_1,p_2\ge 1$ are constants, $b\in L^{p_2}(G)$. Then the Nemytskii operator **f** *is a continuous mapping from $L^{p_1}(G)$ to $L^{p_2}(G)$.*

Proof. From (1.1), we know

$$\int_G |f(x,u(x))|^{p_2}dx \le 2^{p_2}\int_G [a^{p_2}|u(x)|^{p_1}+|b(x)|^{p_2}]dx<\infty,$$

whenever $u\in L^{p_1}(G)$. This implies that $f(x,u(x))\in L^{p_2}(G)$. Thus, **f** is a mapping from $L^{p_1}(G)$ to $L^{p_2}(G)$.

Next, we will show the continuity. Suppose that $\{u_n\}$ converges to u in $L^{p_1}(G)$. It is easy to know that u_n converges to u in measure.

By Lemma 1.2, $f(x, u_n(x))$ converges to $f(x, u(x))$ in measure. From (1.1), it follows

$$
\begin{aligned}
&|f(x, u_n(x)) - f(x, u(x))|^{p_2} \\
&\leq 2^{p_2}(|f(x, u_n(x))|^{p_2} + |f(x, u(x))|^{p_2}) \\
&\leq 4^{p_2}(a^{p_2}|u_n(x)|^{p_1} + a^{p_2}|u(x)|^{p_1} + 2|b(x)|^{p_2}).
\end{aligned}
$$

Therefore, the integral $\int_G |f(x, u_n(x)) - f(x, u(x))|^{p_2} dx$ admits absolute equi-continuity. From the Vitali Theorem, $f(x, u_n(x))$ converges to $f(x, u(x))$ in $L^{p_2}(G)$. This is the end.

It is shown in [13] that the conditions and conclusions in Theorem 1.26 are sufficient and necessary.

Theorem 1.27 (Vitali). *Suppose*

(1) *E is a measurable set and* $\operatorname{mes} E < \infty$.
(2) *$f_1(x), \ldots, f_n(x), \ldots$ are integral functions on E.*
(3) *$f_n(x)$ converges to $f(x)$ in measure.*
(4) *The integrals of $f_n(x)$ admit absolute equi-continuity, that is, for every $\epsilon > 0$, there exists a $\delta > 0$ such that for every A, whenever* $\operatorname{mes} A < \delta$, $|\int_A f_n(x)dx| < \epsilon$ $(n = 1, 2, \ldots)$.

Then $f(x)$ is measurable and

$$\lim_{n\to\infty} \int_E f_n(x)dx = \int_E f(x)dx.$$

1.5.2 Differentiation of Nemytskii Operators

Example 1.8. Let $\Omega \subset \mathbb{R}^n$ be a bounded open domain. Let $C(\bar{\Omega})$ be the Banach space of continuous functions on $\bar{\Omega}$ with the norm $||u|| = \sup_{x\in\bar{\Omega}} |u(x)|$. Let

$$f : \bar{\Omega} \times \mathbb{R} \to \mathbb{R}$$

be a C^1 function and a mapping $F : C(\bar{\Omega}) \to C(\bar{\Omega})$ be defined by

$$u(x) \to f(x, u(x)).$$

Then F is F-differentiable, and $\forall u_0 \in C(\bar{\Omega})$,

$$(F'(u_0) \cdot v)(x) = f_u(x, u_0(x)) \cdot v(x), \quad \forall v \in C(\bar{\Omega}).$$

Proof. For every $h \in C(\bar{\Omega})$ with $||h|| = 1$, we have

$$\frac{1}{t}[F(u_0 + th) - F(u_0)](x) = f_u(x, u_0(x) + t\theta(x)h(x))h(x),$$

where $\theta(x) \in (0,1)$. For $\forall \epsilon > 0$, $\forall M > 0$, there exists a $\delta = \delta(M, \epsilon) > 0$, such that whenever $|\xi|, |\xi'| \leq M$ and $|\xi - \xi'| \leq \delta$, we have

$$|f_u(x, \xi) - f_u(x, \xi')| < \epsilon, \quad \forall x \in \bar{\Omega}.$$

Let $M = ||u_0|| + ||h||$. For any $|t| < \delta < 1$, we obtain

$$|f_u(x, u_0(x) + t\theta(x)h(x)) - f_u(x, u_0(x))| < \epsilon.$$

Hence,

$$DF(u_0, h)(x) = f_u(x, u_0(x))h(x).$$

Note that the multiplication operator $h \to A(u)h = f_u(x, u(x))h(x)$ is continuous linear. The mapping $u \to A(u)$ is continuous from $C(\bar{\Omega})$ to $L(C(\bar{\Omega}), C(\bar{\Omega}))$. Therefore, F is F-differentiable, and

$$(F'(u_0)v)(x) = f_u(x, u_0(x)) \cdot v(x), \quad v \in C(\bar{\Omega}).$$

Example 1.9. Let $X = L^2([0,1])$. Define $F : X \to X$ by

$$F(g)(t) := \sin g(t).$$

Then F is Lipschitz, but is not F-differentiable.

Proof. From

$$\begin{aligned} ||F(g_1) - F(g_2)||^2 &= \int_0^1 |\sin g_1(t) - \sin g_2(t)|^2 dt \\ &\leq \int_0^1 |g_1(t) - g_2(t)|^2 dt = ||g_1 - g_2||^2, \end{aligned}$$

it follows that F is Lipschitz.

Fix $g_0 \in X$, $\forall h \in X$. Firstly, we show that F admits G-derivative. Because

$$\begin{aligned}F(g_0+sh)(t)-F(g_0)(t) &= \sin(g_0(t)+sh(t))-\sin g_0(t)\\ &= \cos(g_0(t)+s\theta(t)h(t))sh(t),\\ &\quad 0\le\theta(t)\le 1,\end{aligned}$$

we claim that $DF(g_0)h(t)=(\cos g_0(t))h(t)$, that is, we will show

$$\int_0^1\left|\frac{\sin(g_0(t)+sh(t))-\sin g_0(t)}{s}-(\cos g_0(t))h(t)\right|^2 dt\to 0\quad (s\to 0),$$

which is equivalent to proving

$$\int_0^1\left|\frac{1}{s}\int_0^s\cos(g_0(t)+\tau h(t))d\tau-\cos g_0(t)\right|^2 h^2(t)dt\to 0\quad (s\to 0).$$

Since the continuous functions space $C[0,1]$ is dense in $L^2[0,1]$, we can assume that h is a continuous square-integrable function. Letting

$$\psi_s(t)=\left|\frac{1}{s}\int_0^s\cos(g_0(t)+\tau h(t))d\tau-\cos g_0(t)\right|^2 h^2(t),\quad s>0,$$

we have $\psi_s(t)\le 4|h(t)|^2$. Since the function $t\mapsto 4|h(t)|^2$ is integrable, we obtain that the functions $t\mapsto\psi_s(t)$ are dominated by a integrable function. For any fixed t, it is transformed to show

$$\lim_{s\to 0}\psi_s(t)=0.$$

In fact, it is known that $|h(t)|<\infty$. Letting $\psi_s(t)=\phi_s^2(t)h^2(t)$, where

$$\begin{aligned}\phi_s(t) &= \left|\frac{1}{s}\int_0^s\cos(g_0(t)+\tau h(t))d\tau-\cos g_0(t)\right|\\ &= \left|\frac{1}{s}\int_0^s[\cos(g_0(t)+\tau h(t))-\cos g_0(t)]d\tau\right|\\ &= \left|\frac{1}{s}\int_0^s(-2)\sin\left(g_0(t)+\frac{\tau h(t)}{2}\right)\sin\frac{\tau h(t)}{2}d\tau\right|\\ &\le \frac{2}{s}\int_0^s\left|\sin\frac{\tau h(t)}{2}\right|d\tau\quad (s\ll 1),\end{aligned}$$

we obtain $\phi_s(t)\to 0\,(s\to 0)$.

Next, we will show that F is not F-differentiable. We prove it by contradiction. If F would be F-differentiable, we would have

$$\lim_{||h||\to 0} \frac{||F(g_0+h)-F(g_0)-DF(g_0)h||}{||h||}=0,$$

that is,

$$\frac{\left(\int_0^1[\sin(g_0(t)+h(t))-\sin g_0(t)-(\cos g_0(t))h(t)]^2dt\right)^{\frac{1}{2}}}{\left(\int_0^1[h(t)]^2dt\right)^{\frac{1}{2}}}\to 0,$$

whenever $(\int_0^1[h(t)]^2dt)^{\frac{1}{2}}\to 0$. Without loss of generality, we assume $g_0\in C[0,1]$.

If $g_0(0)\neq\frac{\pi}{2}+2l\pi$, $l\in\mathbb{N}$, we consider a sequence of functions $\{h_n\}_{n=1}^{\infty}\subset L^2([0,1])$, where

$$h_n(t)=\begin{cases}2\pi, & 0\le t<\frac{1}{n},\\ 0, & \frac{1}{n}<t.\end{cases}$$

It is easy to show that

$$||h_n(t)||=\left(\int_0^1[h_n(t)]^2dt\right)^{\frac{1}{2}}=\left(\int_0^{\frac{1}{n}}(2\pi)^2dt\right)^{\frac{1}{2}}=\frac{2\pi}{\sqrt{n}}\to 0\ \ (n\to\infty)$$

and

$$\begin{aligned}&||F(g_0+h_n)-F(g_0)-DF(g_0)h_n||\\ &=\left(\int_0^1[\sin(g_0(t)+h_n(t))-\sin g_0(t)-(\cos g_0(t))h_n(t)]^2dt\right)^{\frac{1}{2}}\\ &=\left(\int_0^{\frac{1}{n}}[2\pi\cos g_0(t)]^2dt\right)^{\frac{1}{2}}.\end{aligned}$$

Thus, we obtain

$$\begin{aligned}\frac{||F(g_0+h_n)-F(g_0)-DF(g_0)h_n||}{||h_n||}&=\left(\frac{\int_0^{\frac{1}{n}}[2\pi\cos g_0(t)]^2dt}{\frac{(2\pi)^2}{n}}\right)^{\frac{1}{2}}\\ &\to|\cos g_0(0)|,\end{aligned}$$

as $n\to\infty$. But $\cos g_0(0)\neq 0$. We get a contradiction.

If $g_0(0) = \frac{\pi}{2} + 2l\pi$, we consider a sequence of functions $\{\tilde{h}_n\}_{n=1}^{\infty} \subset L^2([0,1])$, where

$$\tilde{h}_n(t) = \begin{cases} \pi, & 0 \le t < \frac{1}{n}, \\ 0, & \frac{1}{n} < t. \end{cases}$$

It is easy to show that

$$||\tilde{h}_n(t)|| = \left(\int_0^1 [\tilde{h}_n(t)]^2 dt\right)^{\frac{1}{2}} = \left(\int_0^{\frac{1}{n}} \pi^2 dt\right)^{\frac{1}{2}} = \frac{\pi}{\sqrt{n}} \to 0 \quad (n \to \infty)$$

and

$$\begin{aligned} &||F(g_0 + \tilde{h}_n) - F(g_0) - DF(g_0)\tilde{h}_n|| \\ &= \left(\int_0^1 [\sin(g_0(t) + \tilde{h}_n(t)) - \sin g_0(t) - (\cos g_0(t))\tilde{h}_n(t)]^2 dt\right)^{\frac{1}{2}} \\ &= \left(\int_0^{\frac{1}{n}} [-2\sin g_0(t) - \pi\cos g_0(t)]^2 dt\right)^{\frac{1}{2}}. \end{aligned}$$

Thus, we obtain

$$\begin{aligned} &\frac{||F(g_0 + \tilde{h}_n) - F(g_0) - DF(g_0)\tilde{h}_n||}{||\tilde{h}_n||} \\ &\to \frac{|2\sin g_0(0) + \pi\cos g_0(0)|^{\frac{1}{2}}}{\pi} = \frac{\sqrt{2}}{\pi} \quad (n \to \infty). \end{aligned}$$

We get a contradiction.

1.5.3 A Variational Functional

Put

$$C_T^{\infty} = \{h : \mathbb{R}^1 \to \mathbb{R}^n;\ h(t)\,\text{is infinite-times continuously differentiable and}\ T\text{-periodic}\}.$$

Lemma 1.3. *Suppose that $u, v \in L^1([0,T], \mathbb{R}^n)$ satisfy*

$$\int_0^T \langle u(t), h'(t)\rangle\, dt = -\int_0^T \langle v(t), h(t)\rangle\, dt, \quad \forall h \in C_T^{\infty}. \tag{1.2}$$

Then we obtain that

$$\int_0^T v(t)dt = 0, \tag{1.3}$$

and there exists a constant vector $c \in \mathbb{R}^n$ *such that*

$$u(t) = \int_0^t v(s)ds + c, \tag{1.4}$$

almost everywhere on $[0, T]$.

Proof. Let $\{e_j\}$ be the standard base in $\mathbb{R}^n$. Take $h = e_j$, $1 \le j \le n$. From (1.2), it follows that

$$\int_0^T \langle v(t), e_j \rangle \, dt = 0, \quad j = 1, 2, \ldots, n.$$

Thus, (1.3) holds.

Put

$$w(t) = \int_0^t v(s)ds.$$

It is known that $w \in C([0, T], \mathbb{R}^n)$ is continuous and

$$\begin{aligned} \int_0^T \langle w(t), h'(t) \rangle \, dt &= \int_0^T \left[\int_0^t \langle v(s), h'(t) \rangle \, ds \right] dt \\ &= \int_0^T \left[\int_s^T \langle v(s), h'(t) \rangle \, dt \right] ds \\ &= \int_0^T \langle v(s), h(T) - h(s) \rangle \, ds \\ &= - \int_0^T \langle v(s), h(s) \rangle \, ds. \end{aligned}$$

From (1.2), we obtain

$$\int_0^T \langle u(t) - w(t), h'(t) \rangle \, dt = 0, \quad \forall h \in C_T^\infty.$$

If we take $h(t) = \sin \frac{2\pi kt}{T} e_j$ and $h(t) = \cos \frac{2\pi kt}{T} e_j$, $1 \le j \le n$, $k \in \mathbb{N}\backslash\{0\}$, respectively, from the theory of Fourier series, we know that there exists a constant vector $c \in \mathbb{R}^n$ such that

$$u(t) - w(t) = c, \quad a.e.\ t \in [0, T].$$

This is the end.

Definition 1.5.2. If u, $v \in L^1([0,T], \mathbb{R}^n)$ satisfy (1.2), v is called the **weak derivative** of u, denoted by $\dot{u}$.

Remark. From the lemma mentioned above and the theory of Fourier series, we know that if the weak derivative of u exists, then the weak derivative is unique, and u is equal to the continuous function $w(t) = \int_0^t \dot{u}(s)ds + c$ almost everywhere on $[0, T]$. So, we usually regard $u(t)$ self as a continuous function, and $u(0) = u(T)$. Clearly, if $\dot{u}$ is continuous on $[0, T]$, then $\dot{u}$ is the derivative of u in normal sense.

Assume $1 < p < \infty$. Let the Sobolev space

$$W_T^{1,p} = \{u \in L^p([0,T], \mathbb{R}^n);\ \dot{u} \in L^p([0,T], \mathbb{R}^n)\}$$

with the norm

$$||u|| = \left(\int_0^T |u(t)|^p dt + \int_0^T |\dot{u}(t)|^p dt \right)^{1/p}.$$

It is well known that $W_T^{1,p}$ is a reflexive Banach space and $C_T^\infty \subset W_T^{1,p}$.

As agreed above, if $u \in W_T^{1,p}$, then $u(t) = \int_0^t \dot{u}(s)ds + c$, and $u(0) = u(T)$.

Lemma 1.4. *There exists a constant $C > 0$ such that*

$$||u||_\infty \le C||u||_{W_T^{1,p}}, \tag{1.5}$$

for every $u \in W_T^{1,p}$. In addition, if $\int_0^T u(t)dt = 0$, we have

$$||u||_\infty \le C||\dot{u}||_{L^p}. \tag{1.6}$$

Here, $||u||_\infty$ denotes the norm of u in L^∞.

Proof. We can assume $n = 1$ by means of considering components. By the mean valued theorem, there is a $\tau \in (0, T)$ such that

$$\frac{1}{T}\int_0^T u(t)dt = u(\tau).$$

From the Hölder inequality, it follows that

$$\begin{aligned}
|u(t)| &= \left| u(\tau) + \int_\tau^t \dot{u}(s)ds \right| \\
&\le |u(\tau)| + \int_0^T |\dot{u}(s)|ds \\
&\le \frac{1}{T}\left|\int_0^T u(s)ds\right| + T^{\frac{1}{q}}\left(\int_0^T |\dot{u}(s)|^p ds\right)^{\frac{1}{p}} \\
&= \frac{1}{T}\left|\int_0^T u(s)ds\right| + T^{\frac{1}{q}}||\dot{u}||_{L^p}
\end{aligned}$$

for $t \in [0, T]$, where $\frac{1}{p} + \frac{1}{q} = 1$. If $\int_0^T u(s)ds = 0$, (1.6) is valid.

In general, we have

$$\begin{aligned}
|u(t)| &\le \frac{1}{T}\left|\int_0^T u(s)ds\right| + T^{\frac{1}{q}}||\dot{u}||_{L^p} \\
&\le T^{-\frac{1}{p}}||u||_{L^p} + T^{\frac{1}{q}}||\dot{u}||_{L^p} \\
&\le (T^{-\frac{1}{p}} + T^{\frac{1}{q}})||u||_{W_T^{1,p}}.
\end{aligned}$$

This is the end.

Example 1.10. Let $L : [0, T] \times \mathbb{R}^n \times \mathbb{R}^n \to \mathbb{R}$, $(t, x, y) \to L(t, x, y)$ admit the following properties:

(1) for almost everywhere t, $L(t, x, y)$ is continuous in (x, y);
(2) for every (x, y), $L(t, x, y)$ is measurable for t.

Suppose that there exist $a \in C(\mathbb{R}^+, \mathbb{R}^+)$, $b \in L([0, T], \mathbb{R}^+)$ and $c \in L^q([0, T], \mathbb{R}^+)$, $1 < q < \infty$, such that, for almost everywhere

$t \in [0,T]$ and for every $(x,y) \in \mathbb{R}^n \times \mathbb{R}^n$,

$$\begin{aligned}|L(t,x,y)| &\le a(|x|)(b(t)+|y|^p),\\ |L_x(t,x,y)| &\le a(|x|)(b(t)+|y|^p),\\ |L_y(t,x,y)| &\le a(|x|)(c(t)+|y|^{p-1}),\end{aligned} \tag{1.7}$$

where $\frac{1}{p}+\frac{1}{q}=1$. Then the functional

$$f(u)=\int_0^T L(t,u(t),\dot{u}(t))dt$$

is continuously differentiable on $W_T^{1,p}$, and

$$f'(u)h=\int_0^T \left\{\langle L_x(t,u(t),\dot{u}(t)),h(t)\rangle + \left\langle L_y(t,u(t),\dot{u}(t)),\dot{h}(t)\right\rangle\right\} dt. \tag{1.8}$$

In addition, if $u \in W_T^{1,p}$ satisfies $f'(u)=0$, then u is a weak solution of the following Euler–Lagrange equation with the periodic boundary conditions

$$\begin{cases}\dot{L}_y(t,u(t),\dot{u}(t)) = L_x(t,u(t),\dot{u}(t)), \quad a.e.\ t\in[0,T],\\ u(0)-u(T)=L_y(0,u(0),\dot{u}(0))-L_y(T,u(T),\dot{u}(T))=0.\end{cases}$$

In particular, if $L_y(t,x,y)\equiv y$, the periodic boundary condition becomes

$$u(0)-u(T)=\dot{u}(0)-\dot{u}(T)=0.$$

Proof. Letting $u,h\in W_T^{1,p}$, $\lambda\in[-1,1]$, $\lambda\ne 0$, we have

$$\begin{aligned}&\frac{f(u+\lambda h)-f(u)}{\lambda}\\ &\quad=\int_0^T \frac{L(t,u(t)+\lambda h(t),\dot{u}(t)+\lambda\dot{h}(t))-L(t,u(t),\dot{u}(t))}{\lambda}dt\\ &\quad=:\int_0^T \phi(t,\lambda)dt.\end{aligned}$$

By the assumptions, for *a.e.* $t \in [0, T]$, we obtain

$$|\phi(t,\lambda)| = \left| \int_0^1 \langle L_x(t, u(t) + s\lambda h(t), \dot{u}(t) + \lambda \dot{h}(t)), h(t)\rangle ds \right.$$

$$\left. + \int_0^1 \langle L_y(t, u(t), \dot{u}(t) + s\lambda \dot{h}(t)), \dot{h}(t)\rangle ds \right|$$

$$\leq \int_0^1 a(|u(t) + s\lambda h(t)|)(b(t) + |\dot{u}(t) + \lambda \dot{h}(t)|^p)|h(t)|ds$$

$$+ \int_0^1 a(|u(t)|)(c(t) + |\dot{u}(t) + s\lambda \dot{h}(t)|^{p-1})|\dot{h}(t)|ds$$

$$\leq a_0[(b(t) + (|\dot{u}(t)| + |\dot{h}(t)|)^p)|h(t)|$$

$$+ (c(t) + (|\dot{u}(t)| + |\dot{h}(t)|)^{p-1})|\dot{h}(t)|],$$

where $a_0 = \max_{(\lambda,t)\in[-1,1]\times[0,T]} a(|u(t) + \lambda h(t)|)$. Since $b \in L^1$, $(|\dot{u}| + |\dot{h}|)^p \in L^1$, $c \in L^q$, $|\dot{h}| \in L^p$ and h is continuous on $[0, T]$, we get

$$|\phi(t,\lambda)| \leq \psi(t) \in L^1([0,T], \mathbb{R}^+).$$

From the Lebesgue dominated convergence theorem, we obtain

$$\lim_{\lambda\to 0} \frac{f(u + \lambda h) - f(u)}{\lambda}$$

$$= \int_0^T \lim_{\lambda\to 0} \phi(t,\lambda)dt$$

$$= \int_0^T \left[\langle L_x(t, u(t), \dot{u}(t)), h(t)\rangle + \langle L_y(t, u(t), \dot{u}(t)), \dot{h}(t)\rangle\right] dt.$$

Since

$$|L_x(t, u(t), \dot{u}(t))| \leq a(|u(t)|)(b(t) + |\dot{u}(t)|^p) \in L^1,$$

$$|L_y(t, u(t), \dot{u}(t))| \leq a(|u(t)|)(c(t) + |\dot{u}(t)|^{p-1}) \in L^q,$$

from Lemma 1.4, we arrive at

$$\begin{aligned}&\left|\int_0^T [\langle L_x(t,u(t),\dot{u}(t)),h(t)\rangle + \langle L_y(t,u(t),\dot{u}(t)),\dot{h}(t)\rangle]dt\right|\\ &\quad\le C_1||h||_\infty + C_2||\dot{h}||_{L^p}\\ &\quad\le C_3||h||_{W_T^{1,p}}.\end{aligned}$$

Hence, we have proved that f admits the G-derivative $f'(u) \in (W_T^{1,p})^*$ at u, which is detailed by (1.8). From (1.7) and Theorem 1.26, it follows that the mapping $u \mapsto (L_x(\cdot,u,\dot{u}), L_y(\cdot,u,\dot{u}))$ is continuous from $W_T^{1,p}$ to $L^1 \times L^q$. Therefore, $f' : W_T^{1,p} \to (W_T^{1,p})^*$ is continuous. In fact, assuming $u_n \to u$ in $W_T^{1,p}$, if $h \in W_T^{1,p}$ and $||h||_{W^{1,p}} \le 1$, we have

$$\begin{aligned}&\left|\int_0^T \Big\{\langle L_x(t,u_n(t),\dot{u}_n(t)) - L_x(t,u(t),\dot{u}(t)),h(t)\rangle\right.\\ &\qquad\left. + \langle L_y(t,u_n(t),\dot{u}_n(t)) - L_y(t,u(t),\dot{u}(t)),\dot{h}(t)\rangle\Big\}dt\right|\\ &\le \int_0^T |L_x(t,u_n(t),\dot{u}_n(t)) - L_x(t,u(t),\dot{u}(t))|dt \cdot ||h||_\infty\\ &\qquad + \left[\int_0^T |L_y(t,u_n(t),\dot{u}_n(t)) - L_y(t,u(t),\dot{u}(t))|^q dt\right]^{\frac{1}{q}} \cdot ||\dot{h}||_{L^p}\\ &\le \Big\{C\int_0^T |L_x(t,u_n(t),\dot{u}_n(t)) - L_x(t,u(t),\dot{u}(t))|dt\\ &\qquad + \left[\int_0^T |L_y(t,u_n(t),\dot{u}_n(t)) - L_y(t,u(t),\dot{u}(t))|^q dt\right]^{\frac{1}{q}}\Big\}||h||_{W_T^{1,p}}\\ &\to 0.\end{aligned}$$

So, we obtain

$$\sup_{||h||_{W_T^{1,p}}\le 1} |f'(u_n)h - f'(u)h| \to 0,$$

that is, $||f'(u_n) - f'(u)|| \to 0$. Therefore, f is F-differentiable.

Finally, if $u \in W_T^{1,p}$ satisfies $f'(u) = 0$, for every $h \in W_T^{1,p}$, we have

$$\int_0^T \langle L_x(t, u(t), \dot{u}(t)), h(t) \rangle dt = -\int_0^T \langle L_y(t, u(t), \dot{u}(t)), \dot{h}(t) \rangle dt.$$

From Lemma 1.3, we know that $L_y(\cdot, u, \dot{u})$ admits the weak derivative, and satisfies the following

$$\begin{cases} \dot{L}_y(t, u(t), \dot{u}(t)) = L_x(t, u(t), \dot{u}(t)), \quad a.e.\ t \in [0, T], \\ u(0) - u(T) = L_y(0, u(0), \dot{u}(0)) - L_y(T, u(T), \dot{u}(T)) = 0. \end{cases}$$

This completes the proof.

This example provides the following ideas: we can study the existence of periodic solutions for the differential equations $\ddot{u} = \nabla_u F(t, u)$ by means of studying the existence of critical points of functional $f'(u) = 0$, where $L(t, u, y) = \frac{1}{2}y^2 + F(t, u)$.

1.6 Higher-Order Derivatives and Taylor Formulas

Let $f : U \to Y$ be F-differentiable. Thus, for every $x \in U$, $f'(x) \in L(X, Y)$. This means that the derivative mapping f' is a nonlinear mapping from U to $L(X, Y)$. If f' is F-differentiable, the second-order derivative of f at x is defined to be the derivative of $f'(x)$ and f is said to be twice differentiable at x. Thus, we have $f''(x) \in L(X, L(X, Y))$. By the same manner, one defines the kth-order derivative mapping $f^{(k)}$ from U to

$$\underbrace{L(X, L(X, \ldots, L(X}_{k\text{-multi}}, Y)) \cdots).$$

In order to characterize the above mentioned space, we introduce the multi-linear mappings from X to Y as follows.

Definition 1.6.1. A mapping $A : X \times X \times \cdots \times X \to Y$ is said to be a **bounded n-multi-linear mapping**, if

(i) $A(x_1, x_2, \ldots, x_n)$ is linear with respect to each variable $x_i (i = 1, \ldots, n)$;

(ii) there exists a constant M such that

$$\|A(x_1, x_2, \ldots, x_n)\| \leq M\|x_1\| \cdots \|x_n\|,$$

$\forall x_i \in X \ (i = 1, \ldots, n)$.

It is easily shown that every bounded n-multi-linear mapping is continuous.

Let $L(\underbrace{X \times \cdots \times X}_{n\text{-multi}}; Y)$ be the space of all bounded n-multi linear mappings from X to Y. The addition and scalar multiplication are defined in the natural manner. The norm is defined by

$$\|A\| = \sup_{\|x_i\|=1} \|A(x_1, x_2, \ldots, x_n)\|.$$

Then $L(\underbrace{X \times \cdots \times X}_{n\text{-multi}}; Y)$ is a Banach space.

Lemma 1.5. *There exists an isometry and isomorphism*

$$\Xi : L(\underbrace{X \times \cdots \times X}_{n\text{-multi}}; Y) \to L(\underbrace{X, L(X, \ldots, L(X}_{n\text{-multi}}, Y)) \cdots).$$

Proof. We will use the induction. Clearly, the result is valid as $n = 1$. Assume that the result is valid as $n = k$. We consider the case as $n = k + 1$.

For every $A \in L(\underbrace{X \times \cdots \times X}_{k+1\text{-multi}}; Y)$, we define a mapping $B \in L(X, L(\underbrace{X \times \cdots \times X}_{k\text{-multi}}; Y))$ by

$$B(x_1)(x_2, \ldots, x_{k+1}) = A(x_1, \ldots, x_{k+1}).$$

In fact, for each $x_1 \in X$, it is easy to see that the mapping $B(x_1) : \underbrace{X \times \cdots \times X}_{k\text{-multi}} \to Y$ is linear with respect to each variable x_i, $2 \leq i \leq k + 1$ and

$$\begin{aligned}\|B(x_1)(x_2, \ldots, x_{k+1})\| &= \|A(x_1, \ldots, x_{k+1})\| \\ &\leq (\|A\| \cdot \|x_1\|)\|x_2\| \cdots \|x_{k+1}\|.\end{aligned}$$

From the definition, $B(x_1) \in L(\underbrace{X \times \cdots \times X}_{k\text{-multi}}; Y)$ and $||B(x_1)|| \leq ||A|| \cdot ||x_1||$. Clearly, the mapping $B : x_1 \in X \to B(x_1) \in L(\underbrace{X \times \cdots \times X}_{k\text{-multi}}; Y)$ is linear and $||B(x_1)|| \leq ||A|| \cdot ||x_1||$. This implies that $B \in L(X, L(\underbrace{X \times \cdots \times X}_{k\text{-multi}}; Y))$ and $||B|| \leq ||A||$.

Clearly, the mapping

$$\iota : A \in L(\underbrace{X \times \cdots \times X}_{k+1\text{-multi}}; Y) \to B \in L(X, L(\underbrace{X \times \cdots \times X}_{k\text{-multi}}; Y))$$

is linear. It remains to show that the mapping ι is surjective.

In fact, for each $B \in L(X, L(\underbrace{X \times \cdots \times X}_{k\text{-multi}}; Y))$, we define a mapping

$$A(x_1, x_2, \ldots, x_{k+1}) = B(x_1)(x_2, \ldots, x_{k+1}).$$

We easily see that A is linear with respect to each variable x_i, $1 \leq i \leq k+1$ and

$$\begin{aligned} ||A(x_1, \ldots, x_{k+1})|| &= ||B(x_1)(x_2, \ldots, x_{k+1})|| \\ &\leq ||B(x_1)|| \cdot ||x_2|| \cdots ||x_{k+1}|| \\ &\leq ||B|| \cdot ||x_1|| \cdots ||x_{k+1}||. \end{aligned}$$

This shows that $A \in L(\underbrace{X \times \cdots \times X}_{k+1\text{-multi}}; Y)$ and $||A|| \leq ||B||$. Therefore, the mapping ι is isomorphism and isometry.

By the induction, $L(\underbrace{X \times \cdots \times X}_{k\text{-multi}}; Y)$ and $L(\underbrace{X, L(X, \ldots, L(X,}_{k\text{-multi}} Y)) \cdots)$ are isometry and isomorphism. Therefore, $L(\underbrace{X \times \cdots \times X}_{k+1\text{-multi}}; Y)$ and $L(\underbrace{X, L(X, \ldots, L(X,}_{k+1\text{-multi}} Y)) \cdots)$ are isometry and isomorphism. This completes the proof.

From the above, the kth-order derivative mapping is $f^{(k)}: U \to L(\underbrace{X \times \cdots \times X}_{k\text{-multi}}; Y)$. If $f^{(k)}$ is continuous in U, we say that f is k **times continuously differentiable** in U or of **class** C^k in U, denoted by $f \in C^{(k)}(U, Y)$.

Lemma 1.6. *If* $f \in C^k(U, Y)$, $x_0 \in U$, *we have*

$$\frac{\partial^k}{\partial t_1 \cdots \partial t_k} f\left(x_0 + \sum_{i=1}^{k} t_i h_i\right)\Bigg|_{t_1=\cdots=t_k=0} = f^{(k)}(x_0)(h_1, \ldots, h_k),$$

$$\forall h_1, \ldots, h_k \in X.$$

Proof. We use the induction. The result is valid as $k = 1$. Assume that the result is valid as $k = n$. For $k = n + 1$, we have

$$\begin{aligned}
&\frac{\partial^{n+1}}{\partial t_1 \cdots \partial t_n \partial t_{n+1}} f\left(x_0 + \sum_{i=1}^{n+1} t_i h_i\right)\Bigg|_{t_1=\cdots=t_n=t_{n+1}=0} \\
&= \frac{\partial}{\partial t_{n+1}}\left(\frac{\partial^n}{\partial t_1 \cdots \partial t_n} f\left(x_0 + \sum_{i=1}^{n+1} t_i h_i\right)\Bigg|_{t_1=\cdots=t_n=0}\right)\Bigg|_{t_{n+1}=0} \\
&= \frac{\partial}{\partial t_{n+1}}\left(f^{(n)}(x_0 + t_{n+1}h_{n+1})(h_1, \ldots, h_n)\right)\big|_{t_{n+1}=0} \\
&= f^{(n+1)}(x_0)(h_1, \ldots, h_n, h_{n+1}).
\end{aligned}$$

This is the end.

Definition 1.6.2. We say that a multi linear mapping $A \in L(X \times \cdots \times X; Y)$ is n-**symmetric**, if for every permutation $\sigma = (\sigma_1, \ldots, \sigma_n)$ of $(1, 2, \ldots, n)$, we have

$$A(x_1, \ldots, x_n) = A(x_{\sigma_1}, \ldots, x_{\sigma_n}).$$

We denote by $L_s^{(n)}(X, Y)$ the set of all n-symmetric mappings.

Theorem 1.28. *If* $f \in C^k(U,Y)$, $f^{(k)}(x) \in L_s^{(k)}(X,Y)$.

Proof. For each $y^* \in Y^*$ and every permutation $\sigma = (\sigma_1, \ldots, \sigma_k)$ of $(1,2,\ldots,k)$, it follows from Lemma 1.6 that

$$\begin{aligned}
&y^*(f^{(k)}(x_0)(h_1,\ldots,h_k))\\
&= y^*\left(\frac{\partial^k}{\partial t_1\cdots\partial t_k} f\left(x_0 + \sum_{i=1}^k t_i h_i\right)\Bigg|_{t_1=\cdots=t_k=0}\right)\\
&= \frac{\partial^k}{\partial t_1\cdots\partial t_k} y^*\left(f\left(x_0 + \sum_{i=1}^k t_i h_i\right)\right)\Bigg|_{t_1=\cdots=t_k=0}\\
&= \frac{\partial^k}{\partial t_{\sigma_1}\cdots\partial t_{\sigma_k}} y^*\left(f\left(x_0 + \sum_{i=1}^k t_{\sigma_i} h_{\sigma_i}\right)\right)\Bigg|_{t_1=\cdots=t_k=0}\\
&= y^*\left(\frac{\partial^k}{\partial t_{\sigma_1}\cdots\partial t_{\sigma_k}} f\left(x_0 + \sum_{i=1}^k t_{\sigma_i} h_{\sigma_i}\right)\Bigg|_{t_1=\cdots=t_k=0}\right)\\
&= y^*(f^{(k)}(x_0)(h_{\sigma_1},\ldots,h_{\sigma_k})).
\end{aligned}$$

By the arbitrariness of y^*, we obtain

$$f^{(k)}(x_0)(h_1,\ldots,h_k) = f^{(k)}(x_0)(h_{\sigma_1},\ldots,h_{\sigma_k}).$$

This completes the proof.

Theorem 1.29 (Taylor Formula). *Let* $f \in C^n(U,Y)$ *be a given mapping, and a closed segment* $L = \{x_0 + th; 0 \le t \le 1\} \subset U$. *Then*

$$f(x_0+h) = \sum_{k=0}^{n-1} \frac{1}{k!} f^{(k)}(x_0)h^{(k)} + R_n(x_0,h),$$

where $h^{(k)} = (\underbrace{h,\ldots,h}_{k\text{-multi}})$, $R_n(x_0,h) = \int_0^1 \frac{(1-t)^{n-1}}{(n-1)!} f^{(n)}(x_0+th)h^{(n)}dt = \frac{1}{n!} f^{(n)}(x_0)h^{(n)} + o(||h||^n)$.

Proof. For every $y^* \in Y^*$, let $\varphi(t) = y^*(f(x_0+th))$. Then $\varphi^{(k)}(t) = y^*(f^{(k)}(x_0+th)h^{(k)})$, $k = 1, \ldots, n$, and $\varphi^{(n)}(t)$ is continuous in $[0,1]$. By the Taylor Theorem in mathematical analysis, we obtain

$$\varphi(1) = \sum_{k=0}^{n-1} \frac{1}{k!}\varphi^{(k)}(0) + \int_0^1 \frac{(1-t)^{n-1}}{(n-1)!}\varphi^{(n)}(t)dt,$$

that is

$$y^*(f(x_0+h)) = y^*\left(\sum_{k=0}^{n-1} \frac{1}{k!} f^{(k)}(x_0)h^{(k)} + R_n(x_0,h)\right).$$

The Taylor formulas will follow from the arbitrariness of y^*. This is the end.

Example 1.11. Let $X = \mathbb{R}^n, Y = \mathbb{R}$. If $f : X \to Y$ is twice continuously differentiable, then

$$f''(x) = H_f(x) = \left(\frac{\partial^2 f}{\partial x_i \partial x_j}\right).$$

Example 1.12. Consider the integral operator (Urysohn Operator)

$$K(u)(x) = \int_\Omega k(x, y, u(y))dy,$$

where $\Omega \subset \mathbb{R}^n$ is a bounded closed set, and $k(x,y,u), k'_u(x,y,u)$, $k''_{uu}(x,y,u)$ are continuous on $\Omega \times \Omega \times \mathbb{R}$. Then $K : C(\Omega) \to C(\Omega)$ admits the second continuous Fréchet derivative, and

$$K''(u)h_1h_2(x) = \int_\Omega k''_{uu}(x,y,u(y))h_1(y)h_2(y)dy.$$

Proof. If we compute the derivative of

$$K(u+th)(x) = \int_\Omega k(x,y,u(y)+th(y))dy$$

with respect to t and let $t = 0$, we obtain

$$\frac{d}{dt}K(u+th)(x)\Big|_{t=0} = \int_\Omega k'_u(x,y,u(y))h(y)dy.$$

In form we obtain

$$K'(u)h(x) := \int_\Omega k'_u(x,y,u(y))h(y)dy.$$

Clearly, $K'(u) : C(\Omega) \to C(\Omega)$ is a bounded linear operator.

In order to show that K is Fréchet differentiable, it suffices to show that $K'(\cdot) : C(\Omega) \to L(C(\Omega), C(\Omega))$ is continuous.

Let $M := ||u||_C := \sup_{x\in\Omega} |u(x)|$. Since $k'_u(x,y,u)$ and $k''_{uu}(x,y,u)$ are uniformly continuous on $\Omega\times\Omega\times[-M-1, M+1]$, for every $\epsilon > 0$, there exists $\delta > 0$, such that

$$|k'_u(x,y,u_1) - k'_u(x,y,u_2)| \le \frac{\epsilon}{\text{mes }\Omega},$$
$$|k''_{uu}(x,y,u_1) - k''_{uu}(x,y,u_2)| \le \frac{\epsilon}{\text{mes }\Omega}.$$

for all $(x,y) \in \Omega\times\Omega$, whenever $|u_1 - u_2| \le \delta$ and $u_1, u_2 \in [-M-1, M+1]$. Thus, if $\tilde{u} \in C(\Omega)$ and $||\tilde{u} - u||_C < \delta$, we obtain

$$\begin{aligned} ||K'(\tilde{u}) - K'(u)|| &= \sup_{||h||_C\le 1} ||K'(\tilde{u})h - K'(u)h||_C \\ &\le \sup_{x\in\Omega}\int_\Omega |k'_u(x,y,\tilde{u}(y)) - k'_u(x,y,u(y))|dy < \epsilon. \end{aligned}$$

This implies that $K'(\cdot)$ is continuous. Therefore, K is Fréchet differentiable, and

$$K'(u)h(x) := \int_\Omega k'_u(x,y,u(y))h(y)dy.$$

Next, we show that K admits the second Fréchet derivative. Let

$$Qh_1h_2(x) := \int_\Omega k''_{uu}(x,y,u(y))h_1(y)h_2(y)dy.$$

Clearly, $Q \in L(C(\Omega) \times C(\Omega), C(\Omega))$. As $||h_1||_C < \delta$, we have

$$\begin{aligned}
&|K'(u+h_1)h_2(x) - K'(u)h_2(x) - Qh_1h_2(x)| \\
&= \left| \int_\Omega \left\{ \int_0^1 [k''_{uu}(x,y,u(y)+th_1(y)) \right.\right. \\
&\qquad \left.\left. - k''_{uu}(x,y,u(y))]dt \right\} h_1(y)h_2(y)dy \right| \\
&\le \epsilon ||h_1||_C ||h_2||_C.
\end{aligned}$$

From the definition, $K''(u)$ exists and $K''(u) = Q$. This is the end.

1.7 Implicit Function Theorem

1.7.1 Implicit Function Theorem

Let X, Y, Λ be real Banach spaces, and $U \subset X \times \Lambda$ be an open set. Assume that $f : U \to Y$ is continuous and there exists $(x_0, \lambda_0) \in U$ such that

$$f(x_0, \lambda_0) = 0. \tag{1.9}$$

We will study the solvability in a neighborhood of λ_0 and the continuous dependence of the solution on the parameter λ for the following nonlinear equation

$$f(x, \lambda) = 0. \tag{1.10}$$

Theorem 1.30. *Suppose that f is partial F-differentiable with respect to the variable x, and the partial derivative mapping f'_x is continuous at (x_0, λ_0). If $f'_x(x_0, \lambda_0) : X \to Y$ is a regular operator, that is, $f'_x(x_0, \lambda_0) : X \to Y$ is a bijection and admits a bounded inverse, then there exist $r, \delta > 0$ such that whenever $||\lambda - \lambda_0|| < \delta$, the equation (1.10) admits a unique continuous solution $x(\lambda)$ in $||x - x_0|| < r$.*

Proof. The existence of solution $x(\lambda)$ for the equation (1.10) is equivalent to the existence of fixed point $x(\lambda)$ for the following mapping

$$\Phi(x, \lambda) = x - [f'_x(x_0, \lambda_0)]^{-1} f(x, \lambda).$$

Thus, we want to show that there exist $\delta > 0, r > 0$ such that whenever $||\lambda - \lambda_0|| < \delta$, the mapping $\Phi(\cdot, \lambda)$ is a contraction mapping from $B(x_0, r) = \{x \in X;\ ||x - x_0|| \leq r\}$ into itself.

From the assumptions, $[f'_x(x_0, \lambda_0)]^{-1}$ is bounded. Thus, there exists a constant $M > 0$ such that

$$||[f'_x(x_0, \lambda_0)]^{-1}|| \leq M.$$

Since $f'_x(x, \lambda)$ and $f(x_0, \lambda)$ are continuous at (x_0, λ_0) and λ_0 respectively, there are constants $r, \delta > 0$ such that whenever $||x - x_0|| \leq r$ and $||\lambda - \lambda_0|| < \delta$, we have

$$||f'_x(x, \lambda) - f'_x(x_0, \lambda_0)|| < \frac{1}{2M}, \tag{1.11}$$

and

$$||f(x_0, \lambda)|| < \frac{r}{2M}.$$

When $||\lambda - \lambda_0|| < \delta$, for every $x \in B(x_0, r)$, we obtain

$$||\Phi'_x(x, \lambda)|| \leq ||[f'_x(x_0, \lambda_0)]^{-1}|| \cdot ||f'_x(x, \lambda) - f'_x(x_0, \lambda_0)|| < \frac{1}{2}.$$

Thus, we arrive at

$$\begin{aligned} ||\Phi(x, \lambda) - \Phi(x_0, \lambda)|| &\leq \sup_{||x - x_0|| \leq r} ||\Phi'_x(x, \lambda)|| \cdot ||x - x_0|| \\ &< \frac{1}{2}||x - x_0|| \leq \frac{1}{2}r, \end{aligned}$$

and

$$||\Phi(x, \lambda) - x_0|| \leq ||\Phi(x, \lambda) - \Phi(x_0, \lambda)|| + ||\Phi(x_0, \lambda) - x_0|| < r.$$

Therefore, $\Phi(\cdot, \lambda)$ is a contraction mapping from $B(x_0, r)$ into itself. When $||\lambda - \lambda_0|| < \delta$, by the contraction mapping principle, $\Phi(\cdot, \lambda)$ has a unique fixed point $x(\lambda) \in B(x_0, r)$ and $x(\lambda_0) = x_0$.

Next, we will show that $x = x(\lambda)$ is continuous in $||\lambda - \lambda_0|| < \delta$. For any $\lambda_1, \lambda_2 \in \Lambda$ satisfying $||\lambda_i - \lambda_0|| < \delta$, $i = 1, 2$, we have

$$\begin{aligned}
&||x(\lambda_1) - x(\lambda_2)|| \\
&= ||\Phi(x(\lambda_1), \lambda_1) - \Phi(x(\lambda_2), \lambda_2)|| \\
&\leq ||\Phi(x(\lambda_1), \lambda_1) - \Phi(x(\lambda_2), \lambda_1)|| + ||\Phi(x(\lambda_2), \lambda_1) - \Phi(x(\lambda_2), \lambda_2)|| \\
&\leq \frac{1}{2}||x(\lambda_1) - x(\lambda_2)|| + ||\Phi(x(\lambda_2), \lambda_1) - \Phi(x(\lambda_2), \lambda_2)||.
\end{aligned}$$

Thus, we obtain

$$||x(\lambda_1) - x(\lambda_2)|| \leq 2||\Phi(x(\lambda_2), \lambda_1) - \Phi(x(\lambda_2), \lambda_2)||. \tag{1.12}$$

Since $\Phi(x, \lambda)$ is continuous, the right-hand side in (1.12) converges to 0 as $\lambda_1 \to \lambda_2$. Thus, $x(\lambda)$ is continuous in $||\lambda - \lambda_0|| < \delta$. This is the end.

Theorem 1.31 (Differentiability). *Suppose that $f \in C^m(U, Y)$, $m \geq 1$, and the partial derivative $f'_x(x_0, \lambda_0) : X \to Y$ is a regular operator. Then, there exist $r, \delta > 0$ such that whenever $||\lambda - \lambda_0|| < \delta$, the equation* (1.10) *admits a unique continuous solution $x(\lambda)$ in $||x - x_0|| < r$, $x(\lambda)$ possesses continuous derivative mapping of order m, and*

$$x'(\lambda) = -[f'_x(x(\lambda), \lambda)]^{-1} f'_\lambda(x(\lambda), \lambda).$$

Proof. Theorem 1.30 implies the existence, uniqueness and continuity of $x = x(\lambda)$. It suffices to prove the differentiability of $x = x(\lambda)$.

Suppose that the constants r, δ in the proof of Theorem 1.30 also satisfy that if $||\lambda - \lambda_0|| \leq \delta$ and $||x - x_0|| \leq r$,

$$||f'_\lambda(x, \lambda)|| \leq M_1, \quad ||[f'_x(x, \lambda)]^{-1}|| \leq M_2,$$

where M_1, M_2 are positive constants. In fact, it is easy to see that

$$\begin{aligned}
f'_x(x, \lambda) &= f'_x(x_0, \lambda_0) + (f'_x(x, \lambda) - f'_x(x_0, \lambda_0)) \\
&= f'_x(x_0, \lambda_0)\{I + [f'_x(x_0, \lambda_0)]^{-1}(f'_x(x, \lambda) - f'_x(x_0, \lambda_0))\}.
\end{aligned}$$

From (1.11), it follows that the inverse of $f'_x(x, \lambda)$ exists and is uniformly bounded in $||x - x_0|| < r$, $||\lambda - \lambda_0|| < \delta$. Thus, if $\lambda_i \in \Lambda$ and

$||\lambda_i - \lambda_0|| < \delta$, $i = 1, 2$,

$$\begin{aligned} &||x(\lambda_1) - x(\lambda_2) + [f'_x(x(\lambda_2), \lambda_2)]^{-1} f'_\lambda(x(\lambda_2), \lambda_2)(\lambda_1 - \lambda_2)|| \\ &\quad \leq ||[f'_x(x(\lambda_2), \lambda_2)]^{-1}|| \\ &\quad\quad \cdot ||f'_x(x(\lambda_2), \lambda_2)(x(\lambda_1) - x(\lambda_2)) + f'_\lambda(x(\lambda_2), \lambda_2)(\lambda_1 - \lambda_2)|| \\ &\quad \leq M_2 ||f'_x(x(\lambda_2), \lambda_2)(x(\lambda_1) - x(\lambda_2)) + f'_\lambda(x(\lambda_2), \lambda_2)(\lambda_1 - \lambda_2)||. \end{aligned} \tag{1.13}$$

Since $f(x(\lambda_2), \lambda_2) = f(x(\lambda_1), \lambda_1) = 0$, it follows that the above inequality is an infinitesimal of higher order than $||x(\lambda_1) - x(\lambda_2)|| + ||\lambda_1 - \lambda_2||$. By (1.12), we have

$$\begin{aligned} ||x(\lambda_1) - x(\lambda_2)|| &\leq 2M_2 \cdot \sup_{||\lambda - \lambda_0|| < \delta, ||x - x_0|| < r} ||f'_\lambda(x, \lambda)|| \cdot ||\lambda_1 - \lambda_2|| \\ &\leq 2M_2 M_1 ||\lambda_1 - \lambda_2||. \end{aligned}$$

Thus, the right-hand side in (1.13) is $o(||\lambda_1 - \lambda_2||)$. From the definition, we know that $x(\lambda)$ is Fréchet differentiable at λ_2, and

$$x'(\lambda_2) = -[f'_x(x(\lambda_2), \lambda_2)]^{-1} f_\lambda(x(\lambda_2), \lambda_2).$$

Since the inverse operation is analytic, by the chain rule and induction we can prove that $x'(\lambda)$ admits continuous derivative mapping of order $m - 1$. This is the end.

Corollary 1.2. *Suppose that $U \subset X$ is an open set, $x_0 \in U$, $f \in C^m(U, Y)$, $m \geq 1$. If $f'(x_0) : X \to Y$ is a regular operator, then there exists a neighborhood V of x_0 and a neighborhood W of $y_0 = f(x_0)$ such that the mapping $f : V \to W$ is C^m diffeomorphism, that is, $f : V \to W$ is homeomorphism, and f, f^{-1} admit continuous derivative mapping of order m.*

1.7.2 Existence of Solutions of Ordinary Differential Equations

Let $D \subset \mathbb{R} \times \mathbb{R}^n$ and $\Lambda \subset \mathbb{R}^m$ be open sets. Suppose that $f : D \times \Lambda \to \mathbb{R}^n$ is continuous and $D_x f(t, x, \lambda)$ is continuous for $(t, x, \lambda) \in D \times \Lambda$.

We consider the initial value problem

$$\begin{cases} \dot{x} = f(t, x, \lambda), \\ x(\sigma) = \xi, \end{cases} \tag{1.14}$$

where $(\sigma, \xi) \in D$.

Theorem 1.32. *Under the above hypotheses, the initial value problem* (1.14) *possesses a unique solution $x(\sigma, \xi, \lambda)(t)$, which is continuous in $(\sigma, \xi, \lambda, t)$. If $f \in C^k(D\times\Lambda, \mathbb{R}^n)$, then $x(\sigma, \xi, \lambda)(t)$ has continuous derivatives of order k in $(\sigma, \xi, \lambda, t)$ in its domain of definition.*

Proof. If x is a solution of (1.14) defined on $[\sigma - \alpha, \sigma + \alpha]$, set

$$t - \sigma =: \alpha\tau, \quad x(\alpha\tau + \sigma) - \xi := z(\tau).$$

Then z must satisfy

$$\begin{cases} \frac{dz(\tau)}{d\tau} - \alpha f(\alpha\tau + \sigma, \xi + z, \lambda) = 0, \quad -1 \le \tau \le 1, \\ z(0) = 0, \end{cases} \tag{1.15}$$

and vice versa.

Let $X = \{\varphi \in C^1([-1, 1], \mathbb{R}^n);\ \varphi(0) = 0\}$, $Y = C([-1, 1], \mathbb{R}^n)$, $F : \mathbb{R} \times \mathbb{R} \times \mathbb{R}^n \times \Lambda \times X \to Y$, where

$$F(\alpha, \sigma, \xi, \lambda, \varphi)(\tau) = \frac{d\varphi(\tau)}{d\tau} - \alpha f(\alpha\tau + \sigma, \xi + \varphi(\tau), \lambda), \quad -1 \le \tau \le 1.$$

Then,

$$F(0, \sigma_0, \xi_0, \lambda_0, 0) = 0, \quad D_\varphi F(0, \sigma_0, \xi_0, \lambda_0, 0)\psi = \frac{d\psi}{d\tau}, \quad \forall(\sigma_0, \xi_0, \lambda_0).$$

If $\psi \in X$ and $\frac{d\psi}{d\tau} = y \in Y$, then $\psi(\tau) = \int_0^\tau y(s)ds$, and ψ is continuous and linear with respect to y. The Implicit Function Theorem implies the existence of a unique solution $z^*(\alpha, \sigma, \xi, \lambda)(\tau)$ of $F(\alpha, \sigma, \xi, \lambda, z) = 0$ in a neighborhood of $(0, \sigma_0, \xi_0, \lambda_0, 0)$ and this solution function has the same smoothness properties as f. This completes the proof.

1.8 Global Implicit Function Theorem

1.8.1 Global Implicit Function Theorem

Definition 1.8.1. Suppose that X, Y are topological spaces, $\phi : X \to Y$ is continuous. ϕ is called **local homeomorphism**, if, for each $x \in X$, there is a neighborhood U of x and a neighborhood V of $y = \phi(x)$ such that ϕ is a homeomorphism from U onto V.

Definition 1.8.2. If X, Y are topological spaces, a mapping $\phi : X \to Y$ is called **proper** if the inverse image of a compact set is compact.

Lemma 1.7. *Suppose that X, Y are Banach spaces, a mapping $\phi : X \to Y$ is proper and local homeomorphism, $\phi(x_0) = y_0$, and $Q = \{(s,t);\ 0 \leq s,t \leq 1\}$. If $F : Q \to Y$ is continuous, $F(0,0) = y_0$, then there is a unique continuous mapping $G : Q \to X$ such that $G(0,0) = x_0$, $\phi(G(s,t)) = F(s,t)$ for $(s,t) \in Q$.*

For its proof, please refer to [8].

Theorem 1.33 (Global Implicit Function Theorem). *If X, Y are Banach spaces, $\phi : X \to Y$ is proper and local homeomorphism, then ϕ is a global homeomorphism from X onto Y.*

Proof. First, we show that ϕ is surjective, that is, for each $y \in Y$, $\phi(x) - y = 0$ has a solution $x \in X$.

Take a point $x_0 \in X$ and put $y_0 = \phi(x_0)$. Let

$$F(t,x) = \phi(x) - y_0 - t(y - y_0) = 0, \quad 0 \leq t \leq 1$$

and

$$S = \{t \in [0,1];\ F(t,x) = 0 \,\text{has a solution}\, x \in X\}.$$

Clearly, $F(0,x) = \phi(x) - y_0$, $F(1,x) = \phi(x) - y$, S is not empty. What we want to do is to prove that S is an open set and a closed set.

For $t_1 \in S$, that is, there is an $x_1 \in X$ such that $F(t_1, x_1) = 0$, then it follows that $\phi(x_1) = y_0 + t_1(y - y_0)$. Since ϕ is local homeomorphism, it follows that t here exists $\delta_1 > 0$ such that $(t_1 - \delta_1, t_1 + \delta_1) \cap [0,1] \subset S$. This shows that S is an open set.

If $t_n \in S$ and $t_n \to t^*$, it follows that there exists $x_n \in X$ such that $F(t_n, x_n) = 0$, that is, $\phi(x_n) = y_0 + t_n(y - y_0)$. Since ϕ is proper, it follows that $x_n \to x^*$ (if it is necessary, take a subsequence). Thus, $\phi(x^*) = y_0 + t^*(y - y_0)$. Therefore, $t^* \in S$.

From the above mentioned, it is shown that S is an open set and a closed set. Thus, $S = [0, 1]$, that is, $\phi(x) - y = 0$ has a solution $x \in X$. Therefore, ϕ is surjective.

Next, we show that ϕ is injective. We argue it by contradiction. Suppose that there exist $x_0, x_1 \in X$ with $x_0 \neq x_1$ such that $\phi(x_0) = \phi(x_1) := y$. Choose $\bar{x} \in X$ which is distinct from x_0 and x_1. We choose one continuous $x_0(t)$ joining $\bar{x}, x_0$ and another continuous curve $x_1(t)$ joining $\bar{x}, x_1$ $(0 \le t \le 1)$. Then both image curves join $\bar{y} = \phi(\bar{x})$ and y. Let $y_0(t) := \phi(x_0(t))$ and $y_1(t) := \phi(x_1(t))$. Since Y is simply connected, there is a continuous mapping $F : Q \to Y$ such that $F(0, t) = y_0(t), F(1, t) = y_1(t), F(s, 0) = \bar{y}, F(s, 1) = y$, $0 \le s, t \le 1$, where $Q = \{(s, t);\ 0 \le s, t \le 1\}$. From Lemma 1.7, it follows that there is a continuous mapping $G : Q \to X$ such that $\phi \circ G = F$, $G(0, t) = x_0(t), G(1, t) = x_1(t)$. But then $\phi(G(s, 1)) = y$ for $0 \le s \le 1$. This contradicts the fact that ϕ is local homeomorphism. This completes the proof.

1.8.2 The Boundary Value Problems of ODE

Consider the boundary value problem

$$\begin{cases} x''(t) + \psi(x(t)) = f(t), & 0 < t < \pi, \\ x(0) = x(\pi) = 0, \end{cases} \tag{1.16}$$

where $f \in C([0, \pi], \mathbb{R})$, $\psi \in C^1(\mathbb{R}, \mathbb{R})$. Suppose that

$$\psi(0) = 0, \quad m^2 < h \le \psi'(s) \le k < (m+1)^2, \quad s \in \mathbb{R}, \tag{1.17}$$

where $m \ge 0$ is an integer, h, k are constants. The condition (1.17) is called the **nonresonance condition**.

Theorem 1.34. *If ψ satisfies condition* (1.17), *then problem* (1.16) *has a unique solution.*

We recall the following lemmas from elementary differential equations before we prove Theorem 1.34.

Lemma 1.8. *If ρ is measurable, bounded, positive on $[0,\pi]$, then the eigenvalues $\mu_1 < \mu_2 < \cdots$ of*

$$\begin{cases} v'' + \mu\rho v = 0, & 0 < t < \pi, \\ v(0) = v(\pi) = 0 \end{cases}$$

satisfy the following properties:

(i) $0 < \mu_1 < \mu_2 < \cdots, \quad \mu_n \to \infty$ *as* $n \to \infty$;
(ii) *each eigenvalue is simple and the corresponding eigenfunction has* $n-1$ *zeros in* $(0,\pi)$;
(iii) μ_n *is a nonincreasing function of* ρ, *if* $\rho_1(t) < \rho_2(t)$ *a.s. on* $[0,\pi]$, *then the corresponding eigenvalues* μ_n^1, μ_n^2 *satisfy* $\mu_n^1 > \mu_n^2$;
(iv) μ_n *is continuous in* ρ *in the topology of* $L^1[0,\pi]$.

Lemma 1.9. *If x is a solution on (a,b) of*

$$(px')' + g_1 x = 0$$

and y is a solution on (a,b) of

$$(py')' + g_2 y = 0$$

with $p > 0$ continuous, g_1, g_2 integrable functions, $g_2 \geq g_1$ on (a,b), $g_2 > g_1$ on a set of positive measure, then there is a zero of y between any two consecutive zeros of x.

Proof of Theorem 1.34. Let $X = \{x \in C^2([0,\pi],\mathbb{R});\ x(0) = x(\pi) = 0\}$, $Z = C([0,\pi],\mathbb{R})$. We want to show that the mapping

$$T:\ X \to Z, \quad Tx(t) = x''(t) + \psi(x(t))$$

is proper and a local homeomorphism.

Firstly, we will show that if $Tx_n = f_n$ and $\{f_n\} \subset Z$ is bounded, then $\{x_n\}$ is bounded in Z. Suppose on the contrary that $\{x_n\}$ is unbounded. Without loss of generality, we may assume that $|x_n|_Z \to \infty$ $(n \to \infty)$. Denote

$$\omega(s) = \frac{\psi(s)}{s}\ (s \neq 0), \quad \omega(0) = \psi'(0),$$

then $h \leq \omega(s) \leq k$. If we set $g_n = \frac{f_n}{|x_n|_Z}$, $z_n = \frac{x_n}{|x_n|_Z}$, then z_n satisfies

$$z_n'' + \omega(x_n) z_n = g_n. \tag{1.18}$$

Since $-\omega(x_n)z_n + g_n$ is bounded in Z, the sequence $\{z_n''\}$ is bounded in Z and $\{z_n\}$ converges to z^* in $\{z \in C^1([0,\pi],\mathbb{R});\ z(0) = z(\pi) = 0\}$ (we will consider a subsequence of $\{z_n\}$, if it is necessary).

For any $w \in C_0^\infty([0,\pi],\mathbb{R})$, it follows from the equation (1.18) that

$$-\int_0^\pi z_n' w' dt + \int_0^\pi \omega(x_n) z_n w dt = \int_0^\pi g_n w dt. \tag{1.19}$$

Since $\{\omega(x_n)\}$ is bounded, it is precompact in the weak $*$ topology of $L^1([0,\pi],\mathbb{R})$. Thus, there is an $\omega^* \in L^1([0,\pi],\mathbb{R})$, $h \le \omega^* \le k$, a subsequence (we label the same) so that $\omega(x_n) \to \omega^*$ in weak $*$ topology. Taking the limit in (1.19), we obtain

$$-\int_0^\pi z^{*\prime} w' dt + \int_0^\pi \omega^* z^* w dt = 0$$

for all $w \in C_0^\infty([0,\pi],\mathbb{R})$. Thus, z^* is a solution (in the generalized sense) of the equation

$$\begin{cases} y'' + \mu\omega^* y = 0, & 0 < t < \pi, \\ y(0) = y(\pi) = 0 \end{cases} \tag{1.20}$$

for $\mu = 1$. Since $m^2 < h \le \omega^* \le k < (m+1)^2$, Lemma 1.8 implies that the eigenvalues $\{\mu_n\}$ of problem (1.20) satisfy

$$\mu_n h < n^2 < (n+1)^2 < \mu_{n+1} k.$$

Thus, $\mu_n = 1$ does not hold for any n, which is impossible. In fact, if there would be a $\mu_{n_0} = 1$, then

$$h < n_0^2 < (n_0 + 1)^2 < \mu_{n_0+1} k,$$

which implies

$$h < k < n_0^2 < (n_0 + 1)^2 < \mu_{n_0+1} k.$$

We get a contradiction from

$$\mu_{n_0-1} h < (n_0 - 1)^2 < n_0^2 < \mu_{n_0} k = k.$$

Therefore, $\{x_n\}$ is bounded in Z.

Since $\{x_n\}$ is bounded in Z, it follows that $\{x''_n\}$ is bounded in Z. Thus, it follows that there is a subsequence of $\{x_n\}$ converging in $C^1_0([0,\pi],\mathbb{R})$. For this same subsequence, $\{x''_n\}$ converges in Z. Therefore, T is proper.

To show that T is locally one-to-one, it suffices to show that $DT(x)$ is invertible for each $x \in X$. It is equivalent to showing that the problem

$$\begin{cases} y''(t) + \psi'(x(t))y(t) = 0, \quad 0 < t < \pi, \\ y(0) = y(\pi) = 0 \end{cases}$$

has only the trivial solution $y = 0$. This follows from the same discussion which is related with (1.20). This completes the proof.

1.9 Bifurcation Problem

Every problem in applications contains several physical parameters which may vary over certain specified sets. Thus, it is important to understand the qualitative behavior of the systems as the parameters vary. Many problems can be written in the following equation

$$F(x, \lambda) = 0, \tag{1.21}$$

where $(x, \lambda) \in X \times \Lambda$, X, Y, and Λ are Banach spaces, and $F : X \times \Lambda \to Y$ is continuous. Λ is often called the **parameter space**. For each $\lambda \in \Lambda$, let

$$S_\lambda = \{x \in X;\ F(x, \lambda) = 0\},$$

which is called **solution set** of the equation (1.21).

Definition 1.9.1. Suppose $F(0, \lambda) = 0$, $\forall \lambda \in \Lambda$, that is, $0 \in S_\lambda$, $\forall \lambda \in \Lambda$. $(0, \lambda_0)$ is called a **bifurcation point** of equation (1.21), if for any neighborhood U of $(0, \lambda_0)$, there exists $(x, \lambda) \in U$ with $x \in S_\lambda \backslash \{0\}$.

For example, $(0, 0)$ is a bifurcation point of the following algebraic equation

$$x^3 - \lambda x = 0, \quad \lambda \in \mathbb{R}.$$

The reason is the following: $x = 0$ solves the equation for all $\lambda \in \mathbb{R}$; as $\lambda \leq 0$, $x = 0$ is the unique solution; but as $\lambda > 0$, we have two more branches of solution $x = \pm\sqrt{\lambda}$.

Generally, suppose that $(x(\lambda), \lambda)$ solves the equation (1.21), that is, $F(x(\lambda), \lambda) = 0$ for any $\lambda \in \Lambda$, where Λ is a neighborhood of λ_0. Let $x_0 = x(\lambda_0)$. (x_0, λ_0) is called a **bifurcation point** of the equation (1.21), if there exists $(x_n, \lambda_n) \to (x_0, \lambda_0)$ with $x_n \neq x(\lambda_n)$ and $F(x_n, \lambda_n) = 0$. The Implicit Function Theorem immediately implies the following theorem.

Theorem 1.35. *Suppose that $F : U \times \Lambda \to Y$ is continuous, and the partial derivative mapping $F'_x(x, \lambda)$ is continuous. If (x_0, λ_0) is a bifurcation point of the equation (1.21), then $F'_x(x_0, \lambda_0) : X \to Y$ is not regular operator, that is, $F'_x(x_0, \lambda_0) : X \to Y$ does not have a bounded inverse.*

The above mentioned theorem is a necessary condition for the existence of a bifurcation point. There is a counter example to show that it is not sufficient condition. For the sake of discussion, we always assume $F(0, \lambda) = 0$, $\forall \lambda \in \Lambda$.

1.9.1 Lyapunov–Schmidt Reduction

Let X, Y be Banach spaces, $A : X \to Y$ be a linear operator, and $\mathrm{Ker}A$ and $\mathrm{Im}A$ denote its null space in X and range in Y, respectively. The linear operator $A : X \to Y$ is called a **Fredholm operator**, if

(i) $\mathrm{Im}A$ is closed in Y,
(ii) $\dim \mathrm{Ker}A < \infty$,
(iii) $\dim \mathrm{Coker}A = \dim Y/\mathrm{Im}A = \mathrm{codim}\,\mathrm{Im}A < \infty$.

We call

$$\mathrm{ind}A = \mathrm{dimKer}A - \mathrm{dimCoker}A$$

the **Fredholm index** of A.

If $A : X \to Y$ is a completely continuous linear operator, it is well known that $\lambda I - A$ is a Fredholm operator with 0-index for any real number $\lambda \neq 0$.

Suppose that $A : X \to Y$ is a Fredholm operator. Let

$$X_1 = \text{Ker}A, \; Y_1 = \text{Im}A.$$

Then we have the direct sum decompositions:

$$X = X_1 \oplus X_2, \quad Y = Y_1 \oplus Y_2,$$

and the projection operator $P : Y \to Y_1$.

Let $F : U \times \Lambda \to Y$ be continuous, where $U \subset X$ is a neighborhood of 0. Suppose that $F'_x(0, \lambda_0)$ is a Fredholm operator. Let $A = F'_x(0, \lambda_0)$. Then X and Y will have the direct sum decompositions and the projection operator $P : Y \to Y_1$ mentioned above. For any $x \in X$, there exists a unique decomposition:

$$x = x_1 + x_2, \quad x_i \in X_i, \quad i = 1, 2.$$

Thus,

$$F(x, \lambda) = 0 \Longleftrightarrow \begin{cases} PF(x_1 + x_2, \lambda) = 0, \\ (I - P)F(x_1 + x_2, \lambda) = 0. \end{cases}$$

Now, $PF'_x(0, \lambda_0) : X_2 \to Y_1$ is a surjection as well as an injection. According to the Banach theorem, it has a bounded inverse mapping. Since $F(0, \lambda_0) = 0$, it follows from the Implicit Function Theorem that there is a unique mapping

$$u : V_1 \times V \to V_2$$

satisfying

$$PF(x_1 + u(x_1, \lambda), \lambda) = 0, \tag{1.22}$$

where V_i is a neighborhood of 0 in $U \cap X_i$, and V is a neighborhood of λ_0. Thus, to solve (1.21) is equivalent to solving the following equation

$$(I - P)F(x_1 + u(x_1, \lambda), \lambda) = 0. \tag{1.23}$$

The procedure mentioned above is called **Lyapunov–Schmidt reduction**, which reduces an infinite dimensional problem to a finite dimensional problem. (1.23) is also called a **bifurcation equation**.

Lemma 1.10. *Under the above assumptions and symbols, if* $F \in C^p(U \times \Lambda, Y)$, $p \geq 1$, *then we obtain*

$$u(0, \lambda) = 0, \quad u'_{x_1}(0, \lambda_0) = 0.$$

Proof. According to the Implicit Function Theorem, the equation (1.22) admits a unique solution $u(x_1, \lambda)$. Since $F(0, \lambda) = 0$, we obtain $u(0, \lambda) = 0$. Using the Implicit Function Theorem again, we obtain that the partial derivative of (1.22) with respect to x_1 at $(0, \lambda_0)$ is

$$PF'_x(0, \lambda_0)(u'_{x_1}(0, \lambda_0)x_1 + x_1) = 0, \quad \forall x_1 \in X_1 = \mathrm{Ker}A.$$

Since $PF'_x(0, \lambda_0) : X_2 \to Y_1$ has a bounded inverse, we obtain

$$u'_{x_1}(0, \lambda_0)x_1 = 0, \quad \forall x_1 \in X_1,$$

which implies $u'_{x_1}(0, \lambda_0) = 0$. This completes the proof.

1.9.2 Bifurcation Theorem

Theorem 1.36 (Crandall–Rabinowitz). *Suppose that* $F \in C^2(U \times \mathbb{R}, Y)$, $F(0, \lambda) = 0$. *If* $A := F'_x(0, \lambda_0)$ *is a Fredholm operator,* $\mathrm{dimKer}A = \mathrm{dimCoker}A = 1$, *and*

$$F''_{x\lambda}(0, \lambda_0)u_0 \notin \mathrm{Im}F'_x(0, \lambda_0), \tag{1.24}$$

$\forall u_0 \in \mathrm{Ker}F'_x(0, \lambda_0)\backslash\{0\}$, *then* $(0, \lambda_0)$ *is a bifurcation point, and there exists a unique* C^1 *curve* $(\lambda, \psi) : (-\delta, \delta) \to \mathbb{R} \times Z$ *such that*

$$\begin{cases} F(su_0 + \psi(s), \lambda(s)) = 0, \\ \lambda(0) = \lambda_0, \ \psi(0) = \psi'(0) = 0, \end{cases}$$

where $\delta > 0$, *and* Z *is the complement space of* $\mathrm{span}\{u_0\}$ *in* X. *Furthermore, there is a neighborhood of* $(0, \lambda_0)$ *such that*

$$F^{-1}(0) = \{(0, \lambda);\ \lambda \in \mathbb{R}\} \cup \{(su_0 + \psi(s), \lambda(s));\ |s| < \delta\}.$$

Proof. Firstly, we write down the bifurcation equation according to the Lyapunov–Schmidt reduction. By assumption,

$$\mathrm{Ker}F'_x(0, \lambda_0) = \mathrm{span}\{u_0\},$$

and there exists $\phi^* \in Y^*\backslash\{0\}$ such that $\mathrm{Ker}\phi^* = \mathrm{Im}F'_x(0, \lambda_0)$, from the Hahn–Banach theorem. Thus, the bifurcation equation can be rewritten

$$g(s,\lambda) = \langle \phi^*, F(su_0 + u(su_0,\lambda),\lambda)\rangle = 0. \tag{1.25}$$

Clearly, $s = 0$ is the trivial solution of the equation (1.25). We want to look for a nontrivial solution. Noticing

$$g'_s(0,\lambda_0) = \langle \phi^*, F'_x(0,\lambda_0)(u_0 + u'_{x_1}(0,\lambda_0)u_0)\rangle,$$

and $u_0 \in \mathrm{Ker}F'_x(0,\lambda_0)$, we arrive at $g'_s(0,\lambda_0) = 0$ by Lemma 1.10. Therefore, it is impossible to show the existence of solution $s = s(\lambda)$ directly from the Implicit Function Theorem.

We consider λ as a function of s. Noticing $g(0,\lambda) = 0$, we consider a new function

$$h(s,\lambda) = \begin{cases} \frac{1}{s}g(s,\lambda), & \text{as } s \neq 0, \\ g'_s(0,\lambda), & \text{as } s = 0. \end{cases}$$

When $s \neq 0$, the solutions of $h(s,\lambda) = 0$ are the same as the equation (1.25). We assume $h \in C^1$ (we postpone the verification to the end). At this moment, we can know

$$h(0,\lambda_0) = g'_s(0,\lambda_0) = 0,$$

and

$$\begin{aligned} h'_\lambda(0,\lambda_0) &= g''_{s\lambda}(0,\lambda_0) \\ &= \langle \phi^*, F''_{x\lambda}(0,\lambda_0)(u_0 + u'_{x_1}(0,\lambda_0)u_0) \\ &\quad + F'_x(0,\lambda_0)u''_{x_1\lambda}(0,\lambda_0)u_0\rangle \\ &= \langle \phi^*, F''_{x\lambda}(0,\lambda_0)u_0\rangle \\ &\neq 0. \end{aligned}$$

From the Implicit Function Theorem, it follows that there is a C^1 curve $\lambda = \lambda(s)$, $|s| < \delta$, such that

$$h(s,\lambda(s)) = 0, \quad \lambda(0) = \lambda_0.$$

Letting

$$\psi(s) = u(su_0,\lambda(s)),$$

we have

$$\psi(0) = u(0, \lambda_0) = 0, \quad \psi'(0) = u'_{x_1}(0, \lambda_0)u_0 + u'_\lambda(0, \lambda_0)\lambda'(0) = 0,$$

and

$$g(s, \lambda(s)) = \langle \phi^*, F(su_0 + u(su_0, \lambda(s)), \lambda(s)) \rangle = 0,$$

that is,

$$F(su_0 + \psi(s), \lambda(s)) = 0.$$

This is what we want.

Next, we want to show $h \in C^1(B_\eta)$, where $B_\eta = \{(s, \lambda) \in \mathbb{R}^2; |s|^2 + |\lambda - \lambda_0|^2 < \eta^2\}$ for some $\eta > 0$. Indeed, it suffices to show that h is C^1 continuous at $s = 0$.

By definition,

$$\lim_{s \to 0} \frac{1}{s} g(s, \lambda) = g'_s(0, \lambda),$$

which implies the continuity of h. Moreover,

$$\begin{aligned} h'_s(0, \lambda) &= \lim_{s \to 0} \frac{1}{s}[h(s, \lambda) - h(0, \lambda)] \\ &= \lim_{s \to 0} \frac{1}{s^2}[g(s, \lambda) - g(0, \lambda) - g'_s(0, \lambda)s] \\ &= \frac{1}{2} g''_{ss}(0, \lambda). \end{aligned}$$

From

$$\begin{aligned} g(s, \lambda) &= g(0, \lambda) + g'_s(0, \lambda)s + \frac{1}{2!} g''_{ss}(0, \lambda)s^2 + o(s^2), \\ g'_s(s, \lambda) &= g'_s(0, \lambda) + g''_{ss}(0, \lambda)s + o(s), \end{aligned}$$

it follows that

$$\begin{aligned} h'_s(s, \lambda) - h'_s(0, \lambda) &= \frac{1}{s^2}[g'_s(s, \lambda)s - g(s, \lambda) - \frac{1}{2} g''_{ss}(0, \lambda)s^2] = o(1) \\ &\quad (|s| \to 0). \end{aligned}$$

Since $h'_\lambda(0, \lambda) = g''_{s\lambda}(0, \lambda)$, we have

$$h'_\lambda(s, \lambda) - h'_\lambda(0, \lambda) = \frac{1}{s} g'_\lambda(s, \lambda) - g''_{s\lambda}(0, \lambda) \to 0 \quad (s \to 0).$$

This is the end.

Theorem 1.37. *Let Λ, X and Y be Banach spaces. Suppose that $F \in C^2(X \times \Lambda, Y)$ possesses the form*

$$F(x, \lambda) = L(\lambda)x + P(x, \lambda),$$

where $L(\lambda) \in L(X, Y)$, $\forall \lambda \in \Lambda$, and

$$P(0, \lambda) = 0,\ P'_x(0, \lambda) = 0,\ P''_{x\lambda}(0, \lambda_0) = 0, \quad \text{for some } \lambda_0 \in \Lambda.$$

If there exists $u_0 \in \mathrm{Ker}L(\lambda_0)\backslash\{0\}$ and a closed linear subspace $Z \subset X$ such that

$$(z, \lambda) \mapsto L(\lambda_0)z + \lambda L'(\lambda_0)u_0 : Z \times \Lambda \to Y$$

is a linear homeomorphism, then there exists a neighborhood U of $(0, \lambda_0)$, $U \subset (\mathrm{span}\{u_0\} \times Z) \times \Lambda$, $\delta > 0$, and a C^1 mapping $s \mapsto (\phi(s), \lambda(s)) : (-\delta, \delta) \to Z \times \Lambda$, such that

$$F^{-1}(0) \cap U \backslash \{(0, \lambda); \lambda \in \Lambda\} = \{(s(u_0 + \phi(s)), \lambda(s));\ |s| < \delta\},$$
$$(\phi(0), \lambda(0)) = (0, \lambda_0).$$

Proof. Define

$$\Phi(s, z, \lambda) = \begin{cases} \frac{1}{s} F(s(u_0 + z), \lambda), & s \neq 0, \\ F'_x(0, \lambda)(u_0 + z), & s = 0. \end{cases}$$

Firstly, we want to show $\Phi \in C^1((\mathbb{R}^1 \times Z) \times \Lambda, Y)$. Indeed, it suffices to show that Φ is continuously differentiable at $s = 0$.

Since

$$\begin{aligned} &\Phi(s, z, \lambda) - \Phi(0, z, \lambda) \\ &\quad = s^{-1}[F(s(u_0 + z), \lambda) - F(0, \lambda) - F'_x(0, \lambda)s(u_0 + z)] \\ &\quad = s \int_0^1 \int_0^1 F''_{xx}(rts(u_0 + z), \lambda) t dt dr \cdot (u_0 + z)^2, \end{aligned}$$

we have

$$\Phi'_s(0, z, \lambda) = \frac{1}{2} F''_{xx}(0, \lambda)(u_0 + z)^2.$$

As $s \to 0$, we have

$$
\begin{aligned}
&\Phi'_s(s,z,\lambda) - \Phi'_s(0,z,\lambda) \\
&= -s^{-2}\Big[F(s(u_0+z),\lambda) - sF'_x(0,\lambda)(u_0+z) \\
&\quad + \frac{s^2}{2}F''_{xx}(0,\lambda)(u_0+z)^2\Big] \\
&\quad + s^{-1}[F'_x(s(u_0+z),\lambda) - F'_x(0,\lambda)](u_0+z) \\
&= -s^{-2}\Big[s^2\int_0^1\int_0^1 F''_{xx}(rts(u_0+z),\lambda)tdtdr(u_0+z)^2 \\
&\quad + \frac{s^2}{2}F''_{xx}(0,\lambda)(u_0+z)^2\Big] \\
&\quad + s^{-1}[F''_{xx}(0,\lambda)s(u_0+z) + o(s)](u_0+z) \\
&= o(1),
\end{aligned}
$$

and

$$
\begin{aligned}
&\Phi'_\lambda(s,z,\lambda) - \Phi'_\lambda(0,z,\lambda) \\
&= s^{-1}[F'_\lambda(s(u_0+z),\lambda) - F'_\lambda(0,\lambda) - F''_{x\lambda}(0,\lambda)s(u_0+z)] \\
&= \frac{1}{s}\int_0^1 [F''_{x\lambda}(ts(u_0+z),\lambda)s(u_0+z) - F''_{x\lambda}(0,\lambda)s(u_0+z)]dt \\
&= o(1).
\end{aligned}
$$

Thus, $\Phi \in C^1$.

It is easy to know that

$$
\begin{aligned}
\Phi(0,z,\lambda) &= L(\lambda)(u_0+z) + P'_x(0,\lambda)(u_0+z), \\
\Phi(0,0,\lambda_0) &= L(\lambda_0)u_0 + P'_x(0,\lambda_0)u_0 = 0,
\end{aligned}
$$

and

$$
\begin{aligned}
&\Phi'_{(z,\lambda)}(0,0,\lambda_0)(\bar{z},\bar{\lambda}) \\
&= \Phi'_z(0,0,\lambda_0)\bar{z} + \Phi'_\lambda(0,0,\lambda_0)\bar{\lambda} \\
&= L(\lambda_0)\bar{z} + L'(\lambda_0)u_0\cdot\bar{\lambda} + P''_{\lambda x}(0,\lambda_0)u_0\bar{\lambda} + P'_x(0,\lambda_0)\bar{z} \\
&= L(\lambda_0)\bar{z} + \bar{\lambda}L'(\lambda_0)u_0,
\end{aligned}
$$

$\forall(\bar{z}, \bar{\lambda}) \in Z \times \Lambda$. By the assumptions, the last linear operator is a homeomorphism. From the Implicit Function Theorem, it follows that there exists a neighborhood U of $(0, \lambda_0)$ and a unique C^1 curve $s \mapsto (\phi(s), \lambda(s)) \in Z \times \Lambda$, $\forall |s| < \epsilon$, such that

$$(\phi(0), \lambda(0)) = (0, \lambda_0), \quad \Phi(s, \phi(s), \lambda(s)) = 0,$$

that is

$$F(s(u_0 + \phi(s)), \lambda(s)) = 0.$$

This is the end.

1.9.3 Hopf Bifurcation Theorem

Consider the nonlinear autonomous ordinary differential equation

$$\frac{dx}{dt} = f(x, \lambda), \tag{1.26}$$

where $f : \mathbb{R}^N \times \mathbb{R} \to \mathbb{R}^N$ is a smooth mapping, f is independent of t. Suppose $f(0, \lambda) = 0$, that is, $x = 0$ is the stationary solution of (1.26). Let $A(\lambda) = \frac{\partial f(x,\lambda)}{\partial x}\big|_{x=0}$. We make the following assumptions:

(H1) $A(\lambda)$ has a pair of complex eigenvalues $\alpha(\lambda) \pm i\beta(\lambda)$ satisfying $\alpha(0) = 0, \beta(0) = \beta_0 > 0$, and the other eigenvalues of $A(0)$ does not lie on the imaginary axis;

(H2) $\alpha'(0) \neq 0$.

Theorem 1.38 (Hopf Bifurcation Theorem). *If* (H1) *and* (H2) *hold, then the equation* (1.26) *possesses a periodic solution with small amplitude in a neighborhood of* $\lambda = 0$, *where the period of the periodic solution is near* $\frac{2\pi}{\beta_0}$.

Proof. From (H1), the system (1.26) can be rewritten as

$$\frac{dx}{dt} = A(\lambda)x + P(x, \lambda), \tag{1.27}$$

where

$$P(0, \lambda) = P'_x(0, 0) = P''_{x\lambda}(0, 0) = 0.$$

By using a coordinate transformation, the system (1.27) becomes

$$\frac{dx}{dt} = \tilde{A}(\lambda)x + \tilde{P}(x, \lambda), \tag{1.28}$$

where

$$\tilde{A}(\lambda) = \begin{pmatrix} \tilde{B}(\lambda) & 0 \\ 0 & \tilde{C}(\lambda) \end{pmatrix}, \quad \tilde{B}(\lambda) = \begin{pmatrix} \alpha(\lambda) & \beta(\lambda) \\ -\beta(\lambda) & \alpha(\lambda) \end{pmatrix},$$

and $e^{\frac{2\pi}{\beta_0}\tilde{C}(\lambda)} - I$ is an invertible matrix. From (H2), the matrix $\tilde{B}(\lambda)$ can be rewritten as

$$B(\mu) = \begin{pmatrix} \mu & \tilde{\beta}(\mu) \\ -\tilde{\beta}(\mu) & \mu \end{pmatrix}.$$

By changing the time scale, we can assume $\tilde{\beta}(0) = 1$.

Thus, we can consider the following system directly

$$\frac{dx}{dt} = A(\mu)x + P(x, \mu), \tag{1.29}$$

where

$$A(\mu) = \begin{pmatrix} B(\mu) & 0 \\ 0 & C(\mu) \end{pmatrix}, \quad B(\mu) = \begin{pmatrix} \mu & \beta(\mu) \\ -\beta(\mu) & \mu \end{pmatrix},$$

and $e^{2\pi C(\mu)} - I$ is an invertible matrix, and

$$\beta(0) = 1, \quad P(0, \mu) = P'_x(0, 0) = P''_{x\mu}(0, 0) = 0.$$

(1.29) can be seen being from (1.27) by the transformation. Letting $t = \omega\tau$, the system (1.29) becomes

$$\frac{dx}{d\tau} = \omega A(\mu)x + \omega P(x, \mu). \tag{1.30}$$

In the following, we discuss the existence of 2π-periodic solutions for the system (1.30). Let

$$\begin{aligned}
&\Lambda = \mathbb{R}^2, \ \rho = (\omega, \mu) \in \Lambda, \\
&X = C^1_{2\pi}(\mathbb{R}^n) = \{u \in C^1(\mathbb{R}, \mathbb{R}^n); \ u \text{ is } 2\pi\text{-periodic}\}, \\
&Y = C_{2\pi}(\mathbb{R}^n) = \{u \in C(\mathbb{R}, \mathbb{R}^n); \ u \text{ is } 2\pi\text{-periodic}\}
\end{aligned}$$

and

$$\begin{aligned} L(\rho)x &= \frac{dx}{d\tau} - \omega A(\mu)x, \\ \mathbf{P}(x,\rho) &= \omega P(x,\mu), \\ \rho_0 &= (1,0). \end{aligned}$$

Setting

$$u_0 = \begin{pmatrix} \sin\tau \\ \cos\tau \\ 0 \\ \vdots \\ 0 \end{pmatrix}, \quad u_1 = \begin{pmatrix} \cos\tau \\ -\sin\tau \\ 0 \\ \vdots \\ 0 \end{pmatrix},$$

then

$$\mathrm{Ker}L(\rho_0) = \mathrm{span}\{u_0, u_1\}.$$

Define

$$Z = \left\{ z \in X;\ \int_0^{2\pi} z(t)u_j(t)dt = 0,\ j = 0,1 \right\}.$$

It remains to verify the following conditions: for any $y \in Y$, the linear equation

$$L(\rho_0)z + \rho L'(\rho_0)u_0 = \frac{dz}{d\tau} - A(0)z - \omega A(0)u_0 - \mu A'(0)u_0 = y$$

has a unique solution $(\omega, \mu, z) \in \mathbb{R}^2 \times Z$. According to the Fredholm Alternative Theorem, this is equivalent to showing that for every $y \in Y$, there exists a unique $(\omega, \mu) \in \mathbb{R}^2$ such that

$$y + \omega A(0)u_0 + \mu A'(0)u_0 \in \mathrm{Ker}L^*(\rho_0)^{\perp}.$$

Since $\mathrm{Ker}L^*(\rho_0) = \mathrm{Ker}L(\rho_0)$, this is again equivalent to showing

$$\det \begin{pmatrix} \int_0^{2\pi}(A(0)u_0(t))^T u_0(t)dt & \int_0^{2\pi}(A'(0)u_0(t))^T u_0(t)dt \\ \int_0^{2\pi}(A(0)u_0(t))^T u_1(t)dt & \int_0^{2\pi}(A'(0)u_0(t))^T u_1(t)dt \end{pmatrix} \neq 0.$$

A direct computation implies that the determinant is equal to $-4\pi^2$. Thus, all assumptions in Theorem 1.37 are satisfied so that our conclusion is obtained. This is the end.

1.10 Ordered Banach Spaces

1.10.1 Cones and Partial Orders

Definition 1.10.1. Let X be a real valued Banach space, and K be a closed convex subset of X. We say that K is a **closed cone** in X, if it admits the following properties:

(1) if $x \in K$ and $\lambda \geq 0$, then $\lambda x \in K$;
(2) if $x \in K$ and $-x \in K$, then $x = 0$.

We shall always assume implicitly that $K \neq \{0\}$. Given a closed cone $K \subset X$, the partial order relations with respect to K are defined as follows:

$$x \leq y \Longleftrightarrow y - x \in K,$$

$$x < y \Longleftrightarrow x \leq y \quad \text{and} \quad x \neq y,$$

$$x \ll y \Longleftrightarrow x \leq y \quad \text{and} \quad y - x \in \overset{\circ}{K},$$

where $\overset{\circ}{K}$ denotes the nonempty interior.

Theorem 1.39. *Let X be a Banach space and $K \subset X$ a closed cone. Then the partial order $\leq$ with respect to K satisfies the following properties*:

(1) $\leq$ *is a* **partial order** *in X, i.e.,*

$$x \leq x,$$

$$x \leq y, \quad y \leq z \Longrightarrow x \leq z,$$

$$x \leq y, \quad y \leq x \Longrightarrow x = y;$$

(2) $\leq$ *is compatible with the linear structure, i.e.,*

$$x \leq u, \quad y \leq v \Longrightarrow x + y \leq u + v,$$

$$x \leq y, \quad \lambda \geq 0 \Longrightarrow \lambda x \leq \lambda y;$$

(3) $\leq$ *is compatible with the topology on X, i.e.,*

$$x_n \leq y_n, \quad x_n \to x, \quad y_n \to y \Longrightarrow x \leq y.$$

Definition 1.10.2. Let X be a Banach space and $\leq$ a partial order on X. If $\leq$ is compatible with the linear structure and topology on X, we say that $(X, \leq)$ is an **ordered Banach space**.

If K is a closed cone in X, it follows from Theorem 1.39 that the partial order induced by K is compatible with the linear structure and topology on X. Thus, X is an ordered Banach space.

Conversely, if $(X, \leq)$ is an ordered Banach space and we set

$$K := \{x \in X;\ 0 \leq x\},$$

then K is a closed cone in X.

Therefore, the closed cone and partial order correspond to each other. Sometimes we use a closed cone and a partial order together.

In n-dimensional Euclidean space $\mathbb{R}^n$, it is easy to see that the closed set

$$K = \{x \in \mathbb{R}^n;\ x_i \geq 0, 1 \leq i \leq n\}$$

is a cone in $\mathbb{R}^n$. Let $m = (m_1, m_2, \ldots, m_n)$, where $m_i \in \{0, 1\}$. If we set

$$K_m = \{x \in \mathbb{R}^n;\ (-1)^{m_i} x_i \geq 0,\ 1 \leq i \leq n\},$$

then K_m is a cone in $\mathbb{R}^n$. The partial order induced by K_m is denoted as $\leq_m$.

Let $C([a, b], \mathbb{R}^n)$ be the Banach space of all continuous functions from $[a, b]$ to $\mathbb{R}^n$. If we set

$$K = \{x \in C([a, b], \mathbb{R}^n);\ x(t) \geq 0,\ \forall t \in [a, b]\},$$

then K is a closed cone in $C([a, b], \mathbb{R}^n)$.

Definition 1.10.3. Let $(X, \leq)$ be an ordered Banach space. If $u, v \in X$ and $u \leq v$, the subset $\{x \in X;\ u \leq x \leq v\}$ is called an **order-interval** in X, denoted by $\langle u, v \rangle$.

If a subset A in X is contained in an order-interval, we say that A is **order-bounded**.

We shall also keep the usual terminology concerning concepts connected with $\leq$. For example, $\{x_n\}$ is said to be **monotonic** (or **order-monotonic**) if $\{x_n\}$ is either increasing, *i.e.* $x_n \leq x_{n+1}$ for all n, or decreasing, *i.e.* $x_n \geq x_{n+1}$ for all n.

The order-boundedness and the norm-boundedness are two different concepts, here we replace boundedness in the norm by norm-boundedness for explaining the order-boundedness. For example, we consider the Banach space $X = C^1[0,1]$ of continuously differentiable functions from $[0,1]$ to $\mathbb{R}$, where the norm is defined by

$$||x|| = \max_{t\in[0,1]} |x(t)| + \max_{t\in[0,1]} |x'(t)|.$$

It is easy to see that

$$K = \{x \in C^1[0,1];\ x(t) \geq 0\}$$

is a closed cone in X. We take $\{x_n\} \subset X$ as $x_n(t) = t^n$, $0 \leq t \leq 1$. Since $0 \leq x_n(t) \leq 1$ as $0 \leq t \leq 1$, we know that the function sequences $x_n(t) = t^n$ are order-bounded. But, $||x_n|| = 1 + n \to \infty$, which means that $\{x_n\}$ is norm-unbounded.

If we consider the Banach space

$$l^2 = \left\{ \{x_i\};\ \sum_{i=1}^{\infty} |x_i|^2 < \infty \right\}$$

and its closed cone

$$K = \{\{x_i\} \in l^2;\ x_i \geq 0\},$$

we will know that the standard basis $\{e_n\}$ in l^2 is norm-bounded, but not order-bounded.

Definition 1.10.4. Let X be a Banach space and $K \subset X$ a closed cone. K is called **normal**, if every order-interval in X is norm-bounded. K is called **regular**, if every order-monotone sequence which is order-bounded is convergent. K is called **fully regular**, if every order-monotone sequence which is norm-bounded is convergent.

Theorem 1.40. *Let X be a Banach space and $K \subset X$ a closed cone. Then the following statements are equivalent to each other.*

(1) *K is normal.*
(2) *There exists a constant $\delta > 0$ such that $||x + y|| \geq \delta$, for any $x, y \in K$ with $||x|| = ||y|| = 1$.*

(3) *There exists a constant* $M > 0$ *such that* $||x|| \leq M||y||$ *as* $0 \leq x \leq y$.
(4) *If* $x_n \leq y_n \leq z_n$ *and* $x_n \to x^*$, $z_n \to x^*$, *then* $y_n \to x^*$.

Proof. (1) $\Longrightarrow$ (2). We prove it by arguing a contradiction. Assume that $\delta = \frac{1}{n^3}$ and there would exist $x_n, y_n \in K$ such that $||x_n|| = ||y_n|| = 1$, but $||x_n + y_n|| \leq \frac{1}{n^3}$. By the completeness, we know that the series $\sum_{n=1}^{\infty} n(x_n + y_n)$ converges to u_0 in X. Thus, $u_0 \in K$ from the positive homogeneity and closed convexity of K. Put

$$u_0 = \sum_{k=1}^{n-1} k(x_k + y_k) + n(x_n + y_n) + \sum_{k=n+1}^{\infty} k(x_k + y_k).$$

Because $\sum_{k=1}^{n-1} k(x_k + y_k) + \sum_{k=n+1}^{\infty} k(x_k + y_k) \in K$, we obtain $u_0 \geq n(x_n + y_n) \geq nx_n \geq 0$. This means that $\{nx_n\}$ is order-bounded, but $||nx_n|| = n \to \infty$, which implies that $\{nx_n\}$ is norm-unbounded. This contradicts to (1).

(2) $\Longrightarrow$ (3). We prove it by arguing a contradiction. Assume that there would exist $x_n, y_n \in K$, $0 \leq x_n \leq y_n$, but $||x_n|| > n||y_n||$, that is, $\frac{||y_n||}{||x_n||} \leq \frac{1}{n} \to 0$. Set $z_n = \frac{y_n - x_n}{||x_n||}$. Then $z_n \in K$ and $||z_n|| \to 1$. From the condition, it follows that

$$\begin{aligned}\delta &\leq \left\|\frac{x_n}{||x_n||} + \frac{z_n}{||z_n||}\right\| \leq \left\|\frac{x_n}{||x_n||} + z_n\right\| + \left\|\frac{z_n}{||z_n||} - z_n\right\| \\ &= \frac{||y_n||}{||x_n||} + \left\|\frac{z_n}{||z_n||} - z_n\right\| \to 0.\end{aligned}$$

This is a contradiction.

(3) $\Longrightarrow$ (4). If (3) holds, it is easy to see that if $0 \leq u_n \leq v_n$ and $v_n \to 0$, then $u_n \to 0$. Hence, we know that (4) holds.

(4) $\Longrightarrow$ (1). We prove it by arguing a contradiction. Assume that there would exist an order-interval $\langle u, v\rangle$ which is norm-unbounded. Set $w = v - u$. Then the order-interval $\langle 0, w\rangle$ is norm-unbounded. Thus, there exist $x_n \in K$, $0 \leq x_n \leq w$, such that $||x_n|| \to \infty$. Hence, $0 \leq \frac{x_n}{||x_n||} \leq \frac{w}{||x_n||}$, and $||\frac{w}{||x_n||}|| \to 0$. From (4) it follows that $\frac{x_n}{||x_n||} \to 0$. This is a contradiction. This completes the proof.

Theorem 1.41. *Let X be a Banach space and $K \subset X$ a closed cone.*

(1) *If K is regular, then K is normal.*
(2) *If K is fully regular, then K is regular.*

Proof. (1) We prove it by arguing a contradiction. If K would not be normal, it follows from (2) in Theorem 1.40 that for $\delta = \frac{1}{n^2}$, there would exist $x_n, y_n \in K$, $||x_n|| = ||y_n|| = 1$ such that $||x_n + y_n|| \le \frac{1}{n^2}$. Thus, $\sum_{n=1}^{\infty}(x_n + y_n)$ converges absolutely to $u \in K$ in X. Set $z_n = x_1 + x_2 + \cdots + x_n$. Then $z_n \le z_{n+1}$, $n = 1, 2, \ldots$ and $z_n \le (x_1 + \cdots + x_n) + (y_1 + \cdots + y_n) \le u$. Hence, the sequence $\{z_n\}$ is monotone and order-bounded. From the regular assumption on K, $\{z_n\}$ is convergent. But, $||z_{n+1} - z_n|| = ||x_n|| = 1$, which is a contradiction.

(2) In order to show that K is regular, it suffices to show that K is normal, because the order-boundedness implies the norm-boundedness for normal cone.

We prove it by arguing a contradiction. Hence, let K be fully regular and suppose that K is not normal. From (2) in Theorem 1.40, it follows that for $\delta = \frac{1}{2^n}$, there exist $x_n, y_n \in K$, $||x_n|| = ||y_n|| = 1$, such that $||x_n + y_n|| \le \frac{1}{2^n}$. Thus, $\sum_{n=1}^{\infty}(x_n + y_n)$ converges to $u \in K$. Set

$$v_n = \begin{cases} (x_1 + y_1) + \cdots + (x_k + y_k), & \text{if } n = 2k, \\ (x_1 + y_1) + \cdots + (x_k + y_k) + x_{k+1}, & \text{if } n = 2k + 1. \end{cases}$$

Then $v_n \le v_{n+1}$, $n = 1, 2, \ldots$, and

$$||v_n|| \le 1 + \sum_{i=1}^{k} ||x_i + y_i|| < 1 + \sum_{i=1}^{\infty} \frac{1}{2^i} = 2.$$

Hence, the sequence $\{v_n\}$ is monotone and norm-bounded. Because K is fully regular, we obtain that $\{v_n\}$ is convergent. But, $||v_{n+1} - v_n|| = 1$, which is a contradiction. This is the end.

Example 1.13. Let $\bar{\Omega} \subset \mathbb{R}^n$ be a bounded closed domain and $C(\bar{\Omega})$ the Banach space of all continuous functions defined on $\bar{\Omega}$. Set

$$K = \{u \in C(\bar{\Omega});\ u(x) \ge 0,\ \forall x \in \bar{\Omega}\}.$$

Then K is normal, but not regular.

Proof. Clearly, K is normal.

We will show that K is not regular. In fact, we take $x_0 \in \bar{\Omega}$ such that

$$\operatorname{dist}(x_0, \partial\Omega) = \sup_{x \in \bar{\Omega}} \operatorname{dist}(x, \partial\Omega).$$

Let

$$f(x) = \frac{\operatorname{dist}(x, \partial\Omega)}{\operatorname{dist}(x_0, \partial\Omega)}, \quad \forall x \in \Omega.$$

Then $f(x_0) = 1$, and $0 \leq f(x) \leq 1$, $\forall x \in \Omega$. Hence, the sequence $\{f^n\}$ is monotone and order-bounded in $C(\bar{\Omega})$, where $f^n(x) = [f(x)]^n$. But, f^n is not convergent in $C(\bar{\Omega})$.

Example 1.14. Let $X = \{c_0\}$, that is

$$X = \{x = (x_1, x_2, \ldots, x_k, \ldots);\ x_k \in \mathbb{R}, x_k \to 0\ (k \to \infty)\}$$

with the norm

$$||x|| = \sup_{k \geq 0} |x_k|.$$

Denote a closed cone

$$K = \{x \in X;\ x_k \geq 0,\ \forall k\}.$$

Then K is regular, but is not fully regular.

In fact, let $x^{(n)} = (x_1^{(n)}, x_2^{(n)}, \ldots) \in X$, $n \geq 1$ such that $\{x^{(n)}\}$ is order-bounded and increasing, that is, there is a $y \in X$ such that $x^{(n)} \leq y$ and $x^{(n)} \leq x^{(n+1)}$, $\forall n$. Then

$$x^{(n)} \to \left(\sup_n x_1^{(n)}, \sup_n x_2^{(n)}, \ldots\right) \quad (n \to \infty).$$

This implies that K is regular. But, if we take $x^{(n)} = \sum_{i=1}^{n} e_i$, where $e_i = (0, \ldots, 0, 1, 0, \ldots)$, *i.e.* the i-th position is 1 and the rest are 0, then $x^{(n)}$ is increasing and norm-bounded. At this time, $x^{(n)}$ is not convergent. This implies that K is not fully regular.

The following two concepts are useful.

Definition 1.10.5. Let X be a Banach space and $K \subset X$ a closed cone. If the interior $\overset{\circ}{K} \neq \emptyset$, then we say that K is a **solid cone**. If $K - K = X$, that is, $\forall x \in X$, there exist $y, z \in K$ such that $x = y - z$, we say that K is a **reproducing cone**.

Theorem 1.42. *If K is a solid cone, then K is a reproducing cone.*

Proof. Let u_0 be an interior of K. Then there exists a $\rho > 0$ such that the closed ball $\bar{B}(u_0, \rho) = \{x;\ ||x - u_0|| \leq \rho\} \subset K$. For every $x \in X$, we have $u_0 \pm \rho \frac{x}{||x||} \in \bar{B}(u_0, \rho) \subset K$. Thus, $x = y - z$, where $y = \frac{||x||}{2\rho}(u_0 + \rho \frac{x}{||x||}) \in K$, $z = \frac{||x||}{2\rho}(u_0 - \rho \frac{x}{||x||}) \in K$. This is the end.

Example 1.15. Let $X = L^p(\Omega)$, $p \geq 1$, $0 < \text{mes}\,\Omega < \infty$, and

$$K = \{\phi;\ \phi \in L^p(\Omega),\ \phi(x) \geq 0\}.$$

Then K is a reproducing cone, but not a solid cone.

Proof. Every $\phi \in L^p(\Omega)$ can be rewritten as $\phi = \phi_1 - \phi_2$, where

$$\phi_1(x) = \begin{cases} \phi(x), & \text{if } \phi(x) \geq 0, \\ 0, & \text{if } \phi(x) < 0, \end{cases}$$

$$\phi_2(x) = \begin{cases} 0, & \text{if } \phi(x) \geq 0, \\ -\phi(x), & \text{if } \phi(x) < 0, \end{cases}$$

$\phi_1(x), \phi_2(x) \in L^p(\Omega)$. This implies that K is a reproducing cone. But, K has no interior.

In fact, for any fixed $u \in L^p(\Omega)$ with $u \geq 0$ and every $\epsilon > 0$, there exists $\delta > 0$ such that, whenever $0 < \text{mes}\ \Omega_1 < \delta$, we have

$$\int_{\Omega_1} (u + 1)^p(x)dx < \epsilon^p.$$

We denote

$$v(x) = \begin{cases} u(x), & \text{if } x \in \Omega \backslash \Omega_1, \\ -1, & \text{if } x \in \Omega_1. \end{cases}$$

Then $v \in B(u, \epsilon)$. But, $v \notin K$. This implies that u is not an interior in K. Hence, K has no interior.

1.10.2 Positive Linear Functionals and Dual Cone

Definition 1.10.6. Let X be a Banach space and $K \subset X$ a closed cone. A bounded linear functional f defined on X is called the **positive linear functional**, if $f(x) \geq 0$ for every $x \in K$.

We denote by K^* the set of all positive linear functionals defined on X. K^* is called the **dual cone** of K.

Clearly, K^* is a closed and convex subset of the dual space X^* of X, and for every nonnegative real λ and every $f \in K^*$, we have $\lambda f \in K^*$. But, K^* may not be a closed cone, because $K^* \cap (-K^*) = \{0\}$ may be violated.

Example 1.16. Let $X = \mathbb{R}^2$ and $K = \{(x, y);\ x \geq 0, y = 0\}$. Then $K^* = \{(u, v);\ u \geq 0, v \in \mathbb{R}\}$ is not a closed cone of $\mathbb{R}^2 = X^*$.

Theorem 1.43. *Let X be a Banach space and $K \subset X$ a closed cone. If K is reproducing, then K^* is a closed cone of X^*.*

Proof. It suffices to prove that if $f \in K^*$ and $-f \in K^*$, then $f = 0$.

Since K is reproducing, for every $x \in X$, there exist $y, z \in K$ such that $x = y - z$. Because $f, -f \in K^*$, we obtain

$$f(x) = f(y) + (-f)(z) \geq 0,$$
$$-f(x) = f(-x) = f(z) + (-f)(y) \geq 0.$$

Thus, $f(x) = 0$. Since x is arbitrary, we have $f = 0$. This is the end.

Theorem 1.44. *Let X be a Banach space, $K \subset X$ a closed cone, and K^* the dual cone of K. Then*

(i) $K^* \neq \{0\}$.
(ii) $x \in K \Longleftrightarrow f(x) \geq 0, \quad \forall f \in K^*$.
(iii) *For every $x \in K \backslash \{0\}$, there exists $f \in K^*$ such that $f(x) > 0$.*

Proof. (i) Let $x_0 \in X \backslash K$. Then there is an ϵ-neighborhood $B_\epsilon(x_0)$ of x_0 such that $B_\epsilon(x_0) \cap K = \emptyset$. From Theorem 1.7, we know that there are a nonzero bounded linear functional f and a constant r, such that

$$f(x) < r, \quad \forall x \in B_\epsilon(x_0),$$
$$f(x) \geq r, \quad \forall x \in K.$$

Since $\lambda K \subset K$ for $\lambda \geq 0$, we have that whenever $x \in K$ and $\lambda \geq 0$,

$$f(\lambda x) = \lambda f(x) \geq r.$$

If we let λ be large enough or small enough respectively, we obtain $f(x) \geq 0$ and $r \leq 0$. Hence, $f \in K^*$ and $f \neq 0$.

(ii) From the definition, it follows that if $x \in K$, $f(x) \geq 0$ for every $f \in K^*$. Conversely, if $x \notin K$, we know from the proof of (i) that there is an $f \in K^*$, such that $f(x) < r \leq 0$.

(iii) If $x \in K\backslash\{0\}$, then $-x \notin K$. By using the same proof as (i), we obtain that there is an $f \in K^*$ such that $f(-x) < r \leq 0$. Hence, $f(x) > 0$. This is the end.

Theorem 1.45. *Let X be a reflexive Banach space, and $K \subset X$ a closed cone. If K is normal, then K is fully regular.*

Proof. Suppose that $\{x_n\}$ is an order-monotone and norm-bounded sequence. Without lose of generality, we assume that $\{x_n\}$ is order-increasing, that is,

$$x_1 \leq x_2 \leq \cdots \leq x_n \leq x_{n+1} \leq \cdots .$$

In a reflexive Banach space, it is well known that a norm-bounded sequence possesses a weak-convergent subsequence. Thus, $\{x_n\}$ has a subsequence $\{x_{n_k}\}$ which is weak-convergent to x. By using Mazur's separation theorem for convex sets, we know that x belongs to the closed convex hull of $\{x_{n_k}\}$. Then there is a sequence $\{y_j\}$ which is composed of a convex combination of $\{x_{n_k}\}$, such that $\{y_j\}$ converges to x, where

$$y_j = \sum_{i=1}^{k_j} \alpha_i^{(j)} x_{n_i}, \quad \alpha_i^{(j)} \geq 0, \quad \sum_{i=1}^{k_j} \alpha_i^{(j)} = 1, \quad j = 1, 2, \ldots .$$

Since the partial order is compatible with the linear structure, we obtain

$$y_j \leq x_{n_{k_j}}, \quad j = 1, 2, \ldots . \tag{1.31}$$

On the other hand, for every $f \in K^*$, $\{f(x_{n_k})\}$ converges monotonically to $f(x)$, and $f(x) - f(x_{n_k}) = f(x - x_{n_k}) \geq 0$. Thus, $x \geq x_{n_k}$ by Theorem 1.44. Hence, $x \geq x_n$, $\forall n$. From (1.31), we obtain

$$0 \leq x - x_{n_{k_j}} \leq x - y_j.$$

Since $||x - y_j|| \to 0$ and K is normal, $||x - x_{n_{k_j}}|| \to 0$. Because $\{x_n\}$ is monotone and K is normal, we obtain $||x_n - x|| \to 0$. This is the end.

If $\Omega \subset \mathbb{R}^n$ is a bounded measurable set, we know from the above theorem that the closed cone K consisting of all nonnegative functions in $L^p(\Omega)\,(p > 1)$ is fully regular.

1.11 Super- and Sub-Solution Method

Let X be a Banach space, $K \subset X$ a closed cone, and $\leq$ a partial order induced by K. Suppose that $D \subset X$ is a subset, and $F : D \to X$ is a mapping. We consider the following equation

$$x - F(x) = 0. \tag{1.32}$$

Definition 1.11.1. Let $x_0 \in D$. If $F(x_0) \leq x_0$, we say that x_0 is a **super-solution** of (1.32). If $F(x_0) \geq x_0$, we say that x_0 is a **sub-solution** of (1.32).

Definition 1.11.2. If $F(x_1) \leq F(x_2)$ for $x_1 \leq x_2$ and $x_1, x_2 \in D$, we say that F is **increasing**. If $F(x_1) \geq F(x_2)$ for $x_1 \leq x_2$ and $x_1, x_2 \in D$, we say that F is **decreasing**.

Theorem 1.46. *Let X be an ordered Banach space, $D \subset X$ a subset, and $F : D \to X$ be increasing. If there are $x_0, y_0 \in D$ such that $x_0 \leq y_0$, the order-interval $\langle x_0, y_0\rangle \subset D$, and x_0, y_0 are a sub-solution and a super-solution of the equation* (1.32) *respectively, then the equation* (1.32) *admits the minimal solution x^* and the maximal solution y^* in the order-interval $\langle x_0, y_0\rangle$, and $x^* \leq y^*$, if one of the following conditions is satisfied.*

(i) *K is normal and F is compact continuous;*
(ii) *K is regular and F is continuous;*
(iii) *X is reflexive, K is normal and F is continuous or weakly continuous.*

Proof. Let $x_n = F(x_{n-1}) = F^n(x_0)$, $y_n = F(y_{n-1}) = F^n(y_0)$, $n = 1, 2, \ldots$. Since F is increasing and x_0, y_0 are a sub-solution and a super-solution of the equation (1.32) respectively, we have

$$x_0 \leq \cdots \leq x_n \leq x_{n+1} \leq \cdots \leq y_{n+1} \leq y_n \leq \cdots \leq y_0.$$

(i) If K is normal, we obtain that the sequences $\{x_n\}, \{y_n\}$ are norm-bounded. By using the compact continuity of F, we know that

$\{x_n\}, \{y_n\}$ admit convergent subsequences $\{x_{n_k}\}, \{y_{n_k}\}$ respectively. Let $x_{n_k} \to x^*$, $y_{n_k} \to y^*$. Since the order is compatible with the topology on X, we obtain from (ii) in Theorem 1.44 that

$$x_0 \le x_n \le x^* \le y^* \le y_n \le y_0, \quad n = 1, 2, \ldots. \tag{1.33}$$

Since K is normal, we have that $\{x_n\}, \{y_n\}$ converge to x^*, y^*, respectively. From the continuity of F, it follows that x^* and y^* are solutions of the equation (1.32).

(ii) If K is regular, then we obtain that $\{x_n\}$ and $\{y_n\}$ are convergent from the monotonicity and order-boundedness of $\{x_n\}$ and $\{y_n\}$. Let $x_n \to x^*, y_n \to y^*$. Thus, (1.33) holds. By the continuity of F, we know that x^* and y^* are solutions of the equation (1.32).

(iii) If X is reflexive and K is normal, we know that K is fully regular from Theorem 1.45. Since $\{x_n\}$ and $\{y_n\}$ are monotone and norm-bounded, we conclude from the definition that there are x^*, y^* such that $x_n \to x^*, y_n \to y^*$. Since the order is compatible with the topology on X, we obtain from (ii) in Theorem 1.44 that

$$x_0 \le x_n \le x^* \le y^* \le y_n \le y_0.$$

According to the continuity or weak-continuity of F, we have that x^* and y^* are solutions of the equation (1.32).

Finally, if $\bar{x}$ is a solution of the equation (1.32) in the order interval $\langle x_0, y_0\rangle$, that is, $x_0 \le \bar{x} \le y_0$, $F(\bar{x}) = \bar{x}$, we obtain from the monotonicity of F that

$$x_1 = F(x_0) \le F(\bar{x}) = \bar{x} \le F(y_0) = y_1.$$

Repeating the above process, we have

$$x_n = F^n(x_0) \le \bar{x} \le F^n(y_0) = y_n.$$

Since the order is compatible with topology on X, we arrive at

$$x^* \le \bar{x} \le y^*.$$

Hence, x^*, y^* are the minimal solution and the maximal solution of the equation (1.32) in the interval $\langle x_0, y_0\rangle$, respectively. This is the end.

The following theorem provides an approach to find a super-solution and a sub-solution, respectively.

Theorem 1.47. *Let X be a Banach space, $K \subset X$ a solid cone, and $F : X \to X$ a mapping.*

(i) *Suppose that x_0 is a solution of the equation (1.32) (e.g., x_0 is usually a trivial solution 0, $F(0) = 0$), F is Frechét differentiable at x_0, and $F'(x_0)$ admits an eigenvalue $\lambda > 0$ and a corresponding eigenvector $v \in \overset{\circ}{K}$. Then there is an $\epsilon > 0$, such that, whenever $t \in (0, \epsilon)$ and $\lambda < 1$, we have*

$$F(x_0 + tv) \ll x_0 + tv, \quad F(x_0 - tv) \gg x_0 - tv;$$

and whenever $\lambda > 1$, we have

$$F(x_0 + tv) \gg x_0 + tv, \quad F(x_0 - tv) \ll x_0 - tv.$$

(ii) *Suppose that F is asymptotically linear at infinite, that is, there is a bounded linear operator $A : X \to X$, denoted as $A = F'(\infty)$, such that*

$$\lim_{||x|| \to \infty} \frac{F(x) - Ax}{||x||} = 0,$$

and A admits an eigenvalue $\lambda \in (0, 1)$ and a corresponding eigenvector $v \in \overset{\circ}{K}$. Then there exists an $M > 0$, such that, whenever $t \geq M$, we have

$$F(tv) \ll tv.$$

Proof. We may assume that the eigenvector v satisfies $||v|| = 1$.

(i) We only consider the case: $\lambda < 1$.

Since $\lambda < 1$ and $v \in \overset{\circ}{K}$, we obtain $(1-\lambda)v \in \overset{\circ}{K}$. Thus, there exists $\epsilon_0 > 0$, such that, for every $u \in B_{\epsilon_0}(0)$, we have $(1 - \lambda)v + u \in \overset{\circ}{K}$. From the definition of Frechét derivative, it follows that there exists an $\epsilon > 0$, such that, whenever $t \in (0, \epsilon)$, we have

$$||F(x_0 \pm tv) - x_0 \mp \lambda tv|| < t\epsilon_0.$$

Then,

$$F(x_0 + tv) - x_0 - \lambda tv \ll (1 - \lambda)tv,$$
$$F(x_0 - tv) - x_0 + \lambda tv \gg (\lambda - 1)tv.$$

Hence,

$$F(x_0 + tv) \ll x_0 + tv, \quad F(x_0 - tv) \gg x_0 - tv.$$

(ii) Since $(1-\lambda)v \in \overset{\circ}{K}$, we obtain that there exists an $\epsilon_0 > 0$ such that, whenever $u \in X$ and $||u|| \leq \epsilon_0$, we have

$$(1 - \lambda)v + u \in \overset{\circ}{K}.$$

Because $\lim_{t\to\infty} \left|\left|\frac{F(tv)}{t} - \lambda v\right|\right| = 0$, we conclude that there exists an $M > 0$ such that, whenever $t \geq M$, we have

$$\frac{F(tv)}{t} \ll \lambda v + (1 - \lambda)v = v.$$

Thus, $F(tv) \ll tv$. This is the end.

In the following, we want to introduce the Krein–Rutman Theorem, which can be seen as an extension of Perron–Frobenius Theorem. The Krein–Rutman Theorem explains when the operator has positive eigenvalue and eigenvector. Please refer to other books [10, 28] for its proof, so that we omit it.

Definition 1.11.3. Let X be a Banach space, $K \subset X$ a closed cone, and $T : X \to X$ a bounded linear operator. If $TK \subset K$, we say that T is **positive**. If $T(K\backslash\{0\}) \subset \overset{\circ}{K}$, then the operator T is said to be **strongly positive**.

Theorem 1.48 (Krein–Rutman Theorem). *Let X be a Banach space, and $K \subset X$ a reproducing cone. Suppose that the bounded linear operator $T : X \to X$ is positive and compact with spectral radius $r(T) > 0$. Then $r(T)$ is an eigenvalue of T with a positive eigenvector.*

Theorem 1.49 (Krein–Rutman Theorem). *Let X be a Banach space and $K \subset X$ a solid cone, i.e. $\overset{\circ}{K} \neq \emptyset$. Suppose that the bounded linear operator $T : X \to X$ is strongly positive and compact. Then the following conclusions are valid.*

(i) *$r(T) > 0$, $r(T)$ is a simple eigenvalue (algebraic multiplicity is 1) with an eigenvector $v \in \overset{\circ}{K}$.*

(ii) *$|\lambda| < r(T)$ for all eigenvalues $\lambda \neq r(T)$, and the corresponding eigenvector v with λ is not positive, i.e. $v \notin K$.*

(iii) *For $y > 0$, the equation $(\lambda I - T)x = y$ has a unique solution $x \gg 0$ if $\lambda > r(T)$ and no solution in K if $\lambda \leq r(T)$. The equation $r(T)x - Tx = -y$ also has no solution in K.*

(iv) *If $S : X \to X$ is a bounded linear operator and $Sx \geq Tx$ on K, then $r(S) \geq r(T)$; moreover $r(S) > r(T)$ if $Sx \gg Tx$ for $x \in K\backslash\{0\}$.*

If a bounded linear operator $T : X \to X$ is noncompact, an extension of Krein–Rutman Theorem has been obtained by R. D. Nussbaum through the Kuratowski noncompact measure. For a norm-bounded subset $B \subset X$, the **Kuratowski measure** of B, denoted by $\alpha(B)$, is defined as

$$\alpha(B) = \inf\left\{d > 0;\ B = \bigcup_{i=1}^{n} B_i,\ n < \infty, \text{ and diameter } (B_i) \leq d \right.$$
$$\left. \text{for all } 1 \leq i \leq n\right\}.$$

Its properties will be discussed in Chapter 2. A continuous mapping $T : X \to X$ is said to be **k-α-Lipschitz** if $\alpha(T(B)) \leq k\alpha(B)$ for all bounded subsets B, where $k \geq 0$ is a constant. Let

$$\alpha(T) = \inf\{k > 0;\ T \text{ is } k\text{-}\alpha\text{-Lipschitz}\}.$$

We call $\alpha(T)$ the **noncompact measure** of T. It is known that there is the Gelfand's formulation for the **spectral radius** $r(T)$ of bounded linear operator T, that is, $\lim_{n\to\infty} ||T^n||^{\frac{1}{n}} = r(T)$. Similarly, there is a representation of **essential spectral radius** $r_e(T)$ by noncompact measure as follows: $r_e(T) = \lim_{n\to\infty}[\alpha(T^n)]^{\frac{1}{n}}$.

Theorem 1.50 (A generalized Krein–Rutman Theorem). *Let X be a Banach space, $K \subset X$ a reproducing cone, and $T : X \to X$ a bounded linear operator. If T is positive and $r_e(T) < r(T)$, then there is a $v \in \overset{\circ}{K}$ such that $Tv = r(T)v$.*

From the Krein–Rutman Theorem, we easily obtain the following theorem.

Theorem 1.51. *Let X be a Banach space, $K \subset X$ a normal and solid cone.*

(i) *If $F : \langle x_0, y_0 \rangle \to \langle x_0, y_0 \rangle$ is increasing and compact continuous, $x_0 = F(x_0) \ll y_0$, F is Frechét differentiable at x_0, $F'(x_0)$ is strongly positive and $r(F'(x_0)) > 1$, then the equation* (1.32) *has the maximal solution $z \gg x_0$ in $\langle x_0, y_0 \rangle$ and $z = \lim_{n\to\infty} F^n(y_0)$.*
(ii) *Suppose that $F : X \to X$ is increasing and compact continuous, $F(0) = 0$, F admits Frechét derivative $F'(0)$ at 0, $F'(0)$ is strongly positive, and $r(F'(0)) > 1$. If F is asymptotically linear at infinite and $r(F'(\infty)) < 1$, then the equation* (1.32) *admits a solution in K.*

1.12 Mixed Monotone Operators

Definition 1.12.1. Let X be a Banach space, and $K \subset X$ a closed cone. A operator $A : K \times K \to X$ is called **mixed monotone**, if $A(x, y)$ is increasing in x and decreasing in y, respectively.

Theorem 1.52. *Let X be a Banach space, and $K \subset X$ a normal and solid cone. For a class of operators $A = B + D$, i.e. $A(x, y) = B(x, y) + D(x)$, we assume that*

(1) *$B : \overset{\circ}{K} \times \overset{\circ}{K} \to \overset{\circ}{K}$ is mixed monotone operator, and there exists a function $\varphi : (0, 1) \times \overset{\circ}{K} \times \overset{\circ}{K} \to (0, \infty)$ such that*

   (i) *$B(tx, t^{-1}y) \geq \varphi(t, x, y)B(x, y)$, $\varphi(t, x, y) > t$, $\forall x, y \in \overset{\circ}{K}$, $t \in (0, 1)$;*
   (ii) *$\varphi(t, x, y)$ is increasing in x and decreasing in y, respectively;*

(2) *there exist $u_0, v_0 \in \overset{\circ}{K}$ such that $u_0 \leq v_0$, $u_0 \leq A(u_0, v_0)$ and $A(v_0, u_0) \leq v_0$;*
(3) *$D : X \to X$ is a positive linear operator such that $D(\overset{\circ}{K}) \subset \overset{\circ}{K} \cup \{0\}$.*

Then A admits a unique fixed point $x^* \in \langle u_0, v_0\rangle$, i.e. $x^* = A(x^*, x^*)$. In addition, for any initial point $x_0 \in \langle u_0, v_0\rangle$, we have $||x_n - x^*|| \to 0\,(n \to \infty)$, where $x_n = A(x_{n-1}, x_{n-1})$, $n = 1, 2, \ldots$.

Proof. It is easy to see that the operator A is mixed monotone. Since K is a normal and solid cone, thus there exists an $r > 0$ such that

$$A(x,y) \leq A(v_0,u_0) \leq rB(u_0,v_0) \leq rB(x,y), \quad \forall x,y \in \langle u_0,v_0\rangle.$$

Setting $r_0 = \frac{1}{r}$ and $\eta(t,x,y) = \frac{\varphi(t,x,y)}{t} - 1$, $\forall t \in (0,1)$, $\forall x,y \in \langle u_0,v_0\rangle$, we have

$$\begin{aligned} A(tx,t^{-1}y) &= B(tx,t^{-1}y) + D(tx) \\ &\geq t\left[1 + \left(\frac{\varphi(t,x,y)}{t} - 1\right)\right] B(x,y) + tD(x) \\ &\geq tA(x,y) + t\eta(t,x,y)r_0A(x,y) \\ &\geq t[1 + r_0\eta(t,x,y)]A(x,y). \end{aligned}$$

Set

$$\begin{aligned} &u_n = A(u_{n-1},v_{n-1}),\ v_n = A(v_{n-1},u_{n-1}), \quad n = 1,2,\ldots, \\ &t_n = \sup\{t > 0;\ u_n \geq tv_n\}, \quad n = 0,1,\ldots. \end{aligned}$$

We can infer that

$$\begin{aligned} &u_0 \leq u_1 \leq \cdots \leq u_n \leq \cdots \leq v_n \leq \cdots \leq v_1 \leq v_0, \\ &u_n \geq t_nv_n, \quad n = 0,1,\ldots. \end{aligned} \tag{1.34}$$

It is easy to see that

$$u_{n+1} \geq u_n \geq t_nv_n \geq t_nv_{n+1}, \quad n = 0,1,\ldots.$$

Thus there exists an $\epsilon > 0$ such that

$$0 < \epsilon \leq t_1 \leq t_2 \leq \cdots \leq t_n \leq \cdots \leq 1.$$

Set $t^* = \lim_{n\to\infty} t_n$. We claim that $t^* = 1$.

Otherwise, if $t^* \in [\epsilon_0, 1)$, we should have $\eta(t^*, u_0, v_0) > 0$. We consider two cases: (a) there exists an N such that $t_N = t^*$; (b) $t_n < t^*$, $\forall n = 1, 2, \ldots$.

Under the case (a), we can obtain that $t_n = t^*, u_n \geq t^* v_n,\ \forall n > N$, and

$$\begin{aligned} u_{n+1} &= A(u_n, v_n) \geq A(t^* v_n, {t^*}^{-1} u_n) \\ &\geq t^*[1 + r_0 \eta(t^*, v_n, u_n)] v_{n+1}, \quad \forall n > N. \end{aligned}$$

Thus

$$t_{n+1} \geq t^*[1 + r_0 \eta(t^*, v_n, u_n)] \geq t^*[1 + r_0 \eta(t^*, u_0, v_0)] > t^*, \quad \forall n > N.$$

Letting $n \to \infty$, we have

$$t^* \geq t^*[1 + r_0 \eta(t^*, u_0, v_0)] > t^*,$$

which is a contradiction.

Under the case (b), we have that

$$\begin{aligned} u_{n+1} = A(u_n, v_n) \geq A(t_n v_n, t_n^{-1} u_n) &= A\left(\frac{t_n}{t^*} t^* v_n, \frac{t^*}{t_n} \frac{1}{t^*} u_n\right) \\ &\geq \frac{t_n}{t^*}\left[1 + r_0 \eta\left(\frac{t_n}{t^*}, t^* v_n, {t^*}^{-1} u_n\right)\right] A(t^* v_n, {t^*}^{-1} u_n) \\ &\geq \frac{t_n}{t^*} t^*[1 + r_0 \eta(t^*, v_n, u_n)] A(v_n, u_n) \\ &\geq t_n[1 + r_0 \eta(t^*, u_0, v_0)] v_{n+1}. \end{aligned}$$

Thus

$$t_{n+1} \geq t_n[1 + r_0 \eta(t^*, u_0, v_0)] \geq t_n.$$

Letting $n \to \infty$, we have

$$t^* \geq t^*[1 + r_0 \eta(t^*, u_0, v_0)] > t^*,$$

which is a contradiction. Hence $t^* = 1$.

From (1.34), it follows that for any natural number k,

$$0 \leq u_{n+k} - u_n, \quad v_n - v_{n+k} \leq v_n - u_n \leq (1 - t_n) v_1, \quad n = 1, 2, \ldots.$$

Since K is a normal cone, there exist $u^*, v^* \in \langle u_0, v_0 \rangle$ such that

$$\begin{aligned} &\|u_n - u^*\| \to 0, \quad \|v_n - v^*\| \to 0 \quad (n \to \infty), \\ &0 \leq v^* - u^* \leq v_n - u_n \leq (1 - t_n) v_1, \quad n = 1, 2, \ldots. \end{aligned}$$

Thus, $v^* = u^* \in \langle u_0, v_0 \rangle$.

If the operator A has other fixed point $y^* \in \langle u_0, v_0 \rangle$, then

$$u_0 \leq A(y^*, y^*) = y^* \leq v_0.$$

Repeating the above iterative procedure, we have

$$\begin{aligned} &u_n \leq y^* \leq v_n, \\ &0 \leq v^* - y^* \leq v_n - y^* \leq v_n - u_n \leq (1 - t_n)v_1. \end{aligned}$$

It follows from the normality of K that $v^* = y^*$. For $x_0 \in \langle u_0, v_0 \rangle$, we consider the following sequences

$$x_n = A(x_{n-1}, x_{n-1}), \quad n = 1, 2, \ldots.$$

By using the above iterative procedure, we have

$$u_n \leq x_n \leq v_n.$$

Therefore, $||v^* - x_n|| \to 0$ $(n \to \infty)$. This completes the proof.

Exercises

1. For any bounded sequence in l^2, show that the coordinate convergence $\Longleftrightarrow$ weak convergence.
2. Given a family of functions

$$u_\epsilon = \frac{\arctan \frac{x}{\epsilon}}{\arctan \frac{1}{\epsilon}}, \quad \forall \epsilon > 0.$$

   Question: Do they have norm convergence or weak convergence in space $L^2(-1, 1)$? Do they have norm convergence or weak convergence in space $H^1(-1, 1)$?
3. Given a bounded closed subset $G \subset \mathbb{R}^n$, if $f(x, u)$ is continuous in $G \times (-\infty, \infty)$, show that the Nemytskii operator $\mathbf{f} : C(G) \to C(G)$ is continuous and bounded.
4. Let

$$f(x) = \begin{cases} x_1 + x_2 + \frac{x_1^3 x_2}{x_1^4 + x_2^2}, & \text{if}\, x = (x_1, x_2) \neq (0, 0), \\ 0, & \text{if}\, x = (x_1, x_2) = (0, 0). \end{cases}$$

   Show that f is continuous at $(0, 0)$, G-differentiable at $(0, 0)$, and G-derivative is at $(0, 0)$, but not F-differentiable at $(0, 0)$.

5. Let X be a Hilbert space with inner product $\langle \cdot, \cdot \rangle$. Find the derivative of $f(x) = ||x||$, $x \neq 0$.
6. Consider a functional $\phi : C([a,b]) \to \mathbb{R}$, defined by

$$\phi(x) = \int_a^b \int_0^{x(\tau)} f(\tau, s) ds d\tau,$$

where $f : [a,b] \times \mathbb{R} \to \mathbb{R}$ is continuous. Show that ϕ is continuously differentiable, and compute $\phi'(x)$.
7. Let $GL(\mathbb{R}^n)$ be the set of all invertible linear mappings from $\mathbb{R}^n$ to $\mathbb{R}^n$. Suppose that $f : GL(\mathbb{R}^n) \to GL(\mathbb{R}^n)$ is defined by $f(A) = A^{-1}$. Show that f is continuously differentiable, and compute the derivative.
8. Compute the derivative for the following operators.

   (1) $F(h) = \int_0^1 \left(\int_0^t h^2(s) ds \right)^2 dt$,

   (a) $h \in C[0,1]$; (b) $h \in L^2(0,1)$.

   (2) $F(h)(t) = \left(\int_0^t h(s) ds \right)^2$,

   (a) $F : L^1(0,1) \to L^1(0,1)$; (b) $F : C[0,1] \to C[0,1]$;
   (c) $F : C[0,1] \to C^1[0,1]$.

9. Let $\Omega \subset \mathbb{R}^n$. Compute twice differentiation of the following functionals.

   (1) $X = W_0^{1,p}(\Omega, \mathbb{R}^1)$, $2 < p < \infty$, $f(u) = \int_\Omega |\nabla u|^p dx$.
   (2) $X = C_0^1(\bar{\Omega}, \mathbb{R}^n)$, $f(u) = \int_\Omega \det(\nabla u(x)) dx$.
   (3) $X = C_0^1(\bar{\Omega}, \mathbb{R}^1)$, $f(u) = \int_\Omega \sqrt{1 + |\nabla u|^2} dx$.

10. Assume that $B : X_1 \times X_2 \to Y$ is a bounded bilinear mapping, where X_1, X_2, Y are Banach spaces. Let $X = X_1 \times X_2$. For $u \in X$, write $u = (u_1, u_2)$. Show that B is C^∞-differentiable, and for all $u, h, k \in X$,

$$dB(u)h = B(h_1, u_2) + B(u_1, h_2),$$
$$d^2 B(u)kh = B(k_1, h_2) + B(h_1, k_2),$$

and $d^n B(u) = 0$, for $n = 3, 4, \ldots$.

11. Let $\Omega \subset \mathbb{R}^n (n \geq 3)$ be a bounded domain with smooth boundary. Suppose that the function $H : \Omega \times \mathbb{R} \to \mathbb{R}$ satisfies the Caratheodory condition and

$$|H(x, u)| \leq a + b|u|^p, \quad \forall x \in \Omega,\ u \in \mathbb{R},$$

where a, b are positive numbers. Denote $h(x, u) = \int_0^u H(x, s)ds$ and

$$\phi : H_0^1(\Omega) \to \mathbb{R}, \quad \phi(u) = \int_\Omega h(x, u(x))dx.$$

Show that ϕ is continuously differentiable if $1 \leq p \leq n^* = \frac{n+2}{n-2}$; ϕ' is weak continuous if $1 \leq p < n^*$.

12. Let $f : \mathbb{R}^2 \times \mathbb{R} \to \mathbb{R}^2$ be defined by

$$f(x, y, \lambda) = (1 - \lambda)(x, y) - (y^3, x^3).$$

Show that $(0, 0, 1)$ is a bifurcation point of $f(x, y, \lambda) = 0$.

13. Consider $f : \mathbb{R}^2 \times \mathbb{R} \to \mathbb{R}^2$ given by

$$f(x, y, \lambda) = \begin{pmatrix} x \\ y \end{pmatrix} - \lambda \begin{pmatrix} 1 & 0 \\ 0 & 2 \end{pmatrix} \begin{pmatrix} x \\ y \end{pmatrix} + \lambda \begin{pmatrix} xy^2 \\ x^2 y \end{pmatrix}.$$

Compute a bifurcation solution of $f(x, y, \lambda) = 0$, *i.e.* a nontrivial solution bifurcated from a bifurcation point.

14. Let X be a Banach space, $K \subset X$ a solid cone, and K^* its dual cone. Show that $x_0 \in \overset{\circ}{K} \Longleftrightarrow f(x_0) > 0,\ \forall f \in K^* \backslash \{0\}$.

15. Let X be a Banach space, $K \subset X$ a cone, and K^* its dual cone. Then

    (a) K is reproducing iff K^* is normal.
    (b) K is normal iff K^* is reproducing.

16. Let X be a Banach space and $K \subset X$ a cone. Fix $e \in K,\ e \neq 0$ and let

$$X_e = \{x \in X;\ \text{there exists a } \lambda > 0 \text{ such that } -\lambda e \leq x \leq \lambda e\}$$

and

$$||x||_e = \inf\{\lambda > 0;\ -\lambda e \leq x \leq \lambda e\}.$$

$||x||_e$ is called e-norm of $x \in X_e$. $K_e =: K \cap X_e$ is called e-cone. If K is normal, then the following conclusions hold.

(i) $(X_e, ||\cdot||_e)$ is a Banach space and there exists a constant $M > 0$ such that $||x|| \leq M||x||_e$, $\forall x \in X_e$.

(ii) K_e is a normal and solid cone of X_e, and

$$\overset{\circ}{K_e} = \{x \in X;\ \text{there exist } \lambda > \tau > 0 \text{ such that } \tau e \leq x \leq \lambda e\}.$$

(iii) If K is solid and $e \in \overset{\circ}{K}$, then $K_e = K$ and $||\cdot||_e$ is equivalent to $||\cdot||$.

Chapter 2

Topological Degree Theory

Topological degree is an important tool and has many applications in the study of topology and differential equations. Topological degree of continuous mappings in finite-dimensional spaces was established by L. J. Brouwer in 1912, usually referred to as the Brouwer degree. There are several definitions of Brouwer degree, and it has been proven that these definitions are all equivalent. We choose one of them to introduce. Topological degree in infinite-dimensional Banach spaces is established by J. Leray and J. Schauder in 1934 for completely continuous field, referred to as the Leray–Schauder degree. It is also considered to be the beginning of nonlinear functional analysis and has important applications in many aspects of mathematics. This chapter mainly introduces the Brouwer degree and the Leray–Schauder degree. A brief introduction is also given to the degree theory of condensing mappings. Readers interested in the degree theory of A-proper mappings and the S^1 degree theory can refer to [13,16].

2.1 Brouwer Degree

2.1.1 Sard Theorem

Let $\Omega \subset \mathbb{R}^n$ be an open subset, and $f \in C^m(\Omega, \mathbb{R}^n)$, $m \geq 1$. For every $x \in \Omega$, $f'(x) : \mathbb{R}^n \to \mathbb{R}^n$ is a linear operator and

$$f'(x) = \left(\frac{\partial f_j(x)}{\partial x_i} \right)_{n \times n},$$

where $f(x) = (f_1(x), f_2(x), \ldots, f_n(x))$, $x = (x_1, x_2, \ldots, x_n)$.

Definition 2.1.1. $x \in \Omega$ is called a **regular point** of f, if the linear operator $f'(x) : \mathbb{R}^n \to \mathbb{R}^n$ is surjective, i.e. $\det f'(x) \neq 0$; otherwise, $x \in \Omega$ is called a **critical point** of f. A point $y \in \mathbb{R}^n$ will be called a **singular value** (or, **critical value**) of f, if there exists a critical point $x \in \Omega$ such that $f(x) = y$; otherwise, y is called a **regular value** of f.

Theorem 2.1 (Sard). *Let $\Omega \subset \mathbb{R}^n$ be open and $f \in C^1(\Omega, \mathbb{R}^n)$. Denote*

$$E = \{x \in \Omega;\ \det f'(x) = 0\}.$$

Then $f(E)$ is a Lebesgue zero measure set, i.e. the set of all singular values of f is a Lebesgue zero measure set.

Proof. Since an open set Ω in $\mathbb{R}^n$ may be written as a countable union of closed cubes $C_i\ (i = 1, 2, \ldots)$, i.e. $\Omega = \cup_{i=1}^{\infty} C_i$, thus $E = \cup_{i=1}^{\infty}(E \cap C_i)$. Hence, $f(E) = \cup_{i=1}^{\infty} f(E \cap C_i)$. In order to show $\text{mes} f(E) = 0$, it is sufficient to show $\text{mes} f(E \cap C) = 0$ for every closed cube C.

Let l be the side length of C. We divide each side into N equal parts and obtain N^n small closed cubes K, where its diameter is $\delta = \sqrt{n}\frac{l}{N}$.

For every $\epsilon > 0$, by the uniform continuity of $f'(x)$ on C, we then find a sufficiently large N such that, whenever x, x_0 belong to the same closed cube K,

$$||f'(x) - f'(x_0)|| < \epsilon.$$

Therefore, for any $x_1, x_0 \in K$, we have

$$\begin{aligned}
&||f(x_1) - f(x_0) - f'(x_0)(x_1 - x_0)|| \\
&\quad = \left\| \int_0^1 f'(x_0 + t(x_1 - x_0))(x_1 - x_0)dt - \int_0^1 f'(x_0)(x_1 - x_0)dt \right\| \\
&\quad \leq ||x_1 - x_0|| \int_0^1 ||f'(x_0 + t(x_1 - x_0)) - f'(x_0)||dt \\
&\quad \leq \epsilon ||x_1 - x_0|| < \delta\epsilon.
\end{aligned}$$

The geometric significance of this inequality is that the distance between the corresponding points in $f(K) - f(x_0) + f'(x_0)x_0$ and

$f'(x_0)K$ is less than $\epsilon\delta$, where $f(K) - f(x_0) + f'(x_0)x_0$ is a translation of image $f(K)$, and $f'(x_0)K$ is an image of K under a linear mapping $f'(x_0)$.

If $x_0 \in K$ be a critical point of f, then the linear mapping $f'(x_0)$ is not invertible. Thus $f'(x_0)K$ cannot span the whole space $\mathbb{R}^n$, but is included in an $(n-1)$-dimensional linear subspace. Denote $M = \max_{x \in C} ||f'(x)||$. Hence, for any two points $y_1^* = f'(x_0)x_1^*$, $y_2^* = f'(x_0)x_2^*$ in $f'(x_0)K$, here $x_1^*, x_2^* \in K$, we obtain

$$||y_1^* - y_2^*|| \leq ||f'(x_0)|| \cdot ||x_1^* - x_2^*|| \leq \frac{\sqrt{n} l M}{N} = \delta M,$$

which implies that the diameter of $f'(x_0)K$ is less than δM.

Therefore, $f(K)-f(x_0)+f'(x_0)x_0$ is included in an n-dimensional square column, where its height is less than $2\epsilon\delta$ and its bottom is an $(n-1)$-dimensional cube with side length $M\delta + 2\epsilon\delta$, respectively.

According to the translation invariance of Lebesgue measure, we conclude

$$\begin{aligned} \text{mes} f(K) &\leq (M\delta + 2\epsilon\delta)^{n-1} 2\epsilon\delta = (M + 2\epsilon)^{n-1} 2\epsilon\delta^n \\ &= 2(M + 2\epsilon)^{n-1} \epsilon n^{\frac{n}{2}} l^n \frac{1}{N^n} \leq A \frac{\epsilon}{N^n}, \end{aligned}$$

where A is a constant. Since there are at most N^n small cubes containing critical points of f in $E \cap C$, we have

$$\text{mes} f(E \cap C) \leq N^n \text{mes} f(K) \leq A\epsilon.$$

From the arbitrariness of ϵ, it can be concluded that $\text{mes} f(E \cap C) = 0$. This is the end.

2.1.2 Brouwer Degree for C^2 Mappings

In complex analysis, the argument principle tell us the relation between the winding number and the number of zeros:

$$w(f(\Gamma), 0) = \frac{1}{2\pi i} \int_{f(\Gamma)} \frac{dz}{z} = \frac{1}{2\pi i} \int_{\Gamma} \frac{f'(z)}{f(z)} dz = N(f, \Gamma),$$

where Γ is a simply closed curve in the complex plane $\mathbb{C}$, $f(z)$ is analytic in the closed domain enclosed by Γ and is not equal to zero on Γ, $N(f,\Gamma)$ is the number of zeros of $f(z)$ in the regions enclosed by Γ (if the multiplicity of a zero point z_0 is m, then z_0 is counted m times), $w(f(\Gamma),0)$ represents how many times $f(\Gamma)$ winds around 0, that is, if z winds along Γ counterclockwise from z_0 to z_0 one time, the winding number $w(f(\Gamma),0)$ is the number that the corresponding $f(z)$ travels along $f(\Gamma)$ one or more times from $w_0 = f(z_0)$ to $w_0 = f(z_0)$. Thus, we know that if $w(f(\Gamma),0) \neq 0$, $f(z) = 0$ admits at least one solution in the regions enclosed by Γ . We can prove (see [10])

$$\frac{1}{2\pi i}\int_{f(\Gamma)} \frac{dz}{z} = \sum_{k=1}^{p} \operatorname{sign} J_f(z_k),$$

where $f^{-1}(0) = \{z_1, \ldots, z_p\}$. If let $f(z) = (u(x,y), v(x,y))$, $J_f(z_k)$ is Jacobian determinant $\frac{\partial(u,v)}{\partial(x,y)}$ at z_k. In the qualitative study of ordinary differential equations, $P_0 = (x_0, y_0)$ is called a singularity of a plane vector field $V(x,y) = (V_1(x,y), V_2(x,y))$, if $V_1(x_0,y_0) = V_2(x_0,y_0) = 0$. Let C be a closed curve surrounding P_0 in the phase plane. The winding number of a closed curve C is the Poincaré index of a singularity P_0. It is a quantity that characterizes the topological properties of singularities.

In the more general case of continuous maps f from subsets of $\mathbb{R}^n$ into $\mathbb{R}^n$ we also want to define this quantity, called Brouwer degree now, which can show the existence of zeros for the equation $f(x) = 0$. The early statements and discussions on topological degree often relied on algebraic topology methods [14]. In order to facilitate analysts to master this tool, after the efforts of many scholars, it is now known that the normalization, additivity, and homotopy invariance can serve as the axiom for defining degrees. Topological degree can also be established on the basis of analysis.

Let us first define the Brouwer degree from a C^2-continuous function.

Definition 2.1.2. Let $\Omega \subset \mathbb{R}^n$ be a bounded open subset, $f : \bar{\Omega} \to \mathbb{R}^n$ a C^2-continuous, $p \in \mathbb{R}^n$, and $p \notin f(\partial\Omega)$. We define the **Brouwer degree** $\deg(f,\Omega,p)$ of f on Ω with respect to p as follows:

If p is a regular value of f, we define

$$\deg(f,\Omega,p) = \sum_{x\in f^{-1}(p)} \operatorname{sign}\det f'(x). \tag{2.1}$$

If p is a critical value of f, we take a regular value p_1 of f satisfying $||p_1 - p|| < \rho := \text{dist}(p, f(\partial\Omega))$, and define

$$\deg(f, \Omega, p) = \deg(f, \Omega, p_1). \tag{2.2}$$

If p is a regular value of f, from the Implicit Function Theorem, we conclude that

$$f^{-1}(p) = \{x \in \Omega;\ f(x) = p\}$$

is a finite set. Thus, (2.1) is well defined.

For explaining that (2.2) is well defined, we need to show that for any two regular values p_1, p_2 of f satisfying $||p_i - p|| < \rho = \text{dist}(p, f(\partial\Omega))$, $i = 1, 2$,

$$\deg(f, \Omega, p_1) = \deg(f, \Omega, p_2). \tag{2.3}$$

For doing this, we take $\phi \in C^\infty(\mathbb{R}^n)$, such that $\phi \geq 0$, ϕ is equal to 0 outside the ball $B(0, 1)$, and $\int_{\mathbb{R}^n} \phi(x)dx = 1$. Such function is called **mollifier**. A typical example is

$$\phi(x) = \begin{cases} c\exp(\frac{1}{||x||^2 - 1}), & \text{if } ||x|| < 1, \\ 0, & \text{if } ||x|| \geq 1. \end{cases}$$

Let

$$\phi_\epsilon(x) = \frac{1}{\epsilon^n}\phi\left(\frac{x}{\epsilon}\right).$$

Thus, if $||x|| > \epsilon$, $\phi_\epsilon(x) = 0$, and $\int_{\mathbb{R}^n} \phi_\epsilon(x)dx = 1$.

We will convert (2.1) into an integral form as follows.

Lemma 2.1. *Let $\Omega \subset \mathbb{R}^n$ be a bounded open subset, $f \in C^1(\bar{\Omega}, \mathbb{R}^n)$, p be a regular value of f, and $p \notin f(\partial\Omega)$. Then there exists $\epsilon_0 = \epsilon_0(p, f)$ such that, if $0 < \epsilon < \epsilon_0$,*

$$\deg(f, \Omega, p) = \int_\Omega \phi_\epsilon(f(x) - p)J_f(x)dx, \tag{2.4}$$

where $J_f(x) = \det f'(x)$.

Proof. (1) Suppose $f^{-1}(p) = \emptyset$. Let $\epsilon_0 = \mathrm{dist}(p, f(\bar{\Omega})) > 0$. If $0 < \epsilon < \epsilon_0$, $\phi_\epsilon(f(x) - p) \equiv 0$. Thus, the equality (2.4) holds.

(2) Suppose $f^{-1}(p) \neq \emptyset$. Let $f^{-1}(p) = \{x_1, \ldots, x_k\}$. By the Inverse Function Theorem, there exists $\epsilon_0 > 0$ such that

$$f^{-1}(B(p, \epsilon_0)) = \bigcup_{i=1}^{k} U_i,$$

where $U_i \subset \Omega$ is a neighborhood of x_i, U_i and U_j are disjoint each other ($i \neq j$), f restricted on U_i is a diffeomorphism from U_i onto $B(p, \epsilon_0)$, $J_f(x)$ has the same sign on each U_i, and $||f(x) - p|| \geq \epsilon_0$ as $x \in \Omega \backslash \bigcup_{i=1}^{k} U_i$. Thus, if $0 < \epsilon < \epsilon_0$, then

$$\int_\Omega \phi_\epsilon(f(x) - p) J_f(x) dx = \sum_{i=1}^{k} \mathrm{sign} J_f(x_i) \int_{U_i} \phi_\epsilon(f(x) - p) |J_f(x)| dx. \tag{2.5}$$

By using variable substitution $y = f(x)$, we have $dy = |J_f(x)| dx$ and

$$\begin{aligned} \int_{U_i} \phi_\epsilon(f(x) - p) |J_f(x)| dx &= \int_{B(p,\epsilon_0)} \phi_\epsilon(y - p) dy \\ &= \int_{B(0,\epsilon_0)} \phi_\epsilon(y) dy = 1. \end{aligned}$$

Combined with (2.5), we obtain

$$\deg(f, \Omega, p) = \int_\Omega \phi_\epsilon(f(x) - p) J_f(x) dx.$$

This is the end.

Remark. (2.4) is independent of the mollifier ϕ.

Hence, in order to show the rationality of (2.2), it is sufficient to show

$$\int_\Omega [\phi_\epsilon(f(x) - p_1) - \phi_\epsilon(f(x) - p_2)] J_f(x) dx = 0. \tag{2.6}$$

Lemma 2.2. *Let*

$$G(x) = (p_1 - p_2) \int_0^1 \phi_\epsilon(x - p_1 + t(p_1 - p_2))dt.$$

Then $\text{div}G(x) = \phi_\epsilon(x - p_2) - \phi_\epsilon(x - p_1)$. *In addition, letting* $\delta = \rho - \max\{||p_1 - p||, ||p_2 - p||\}$ *and* $0 < \epsilon < \min\{\delta, \epsilon_0\}$, $G(f(x))$ *is identically equal to zero on* $\partial\Omega$, *where* ϵ_0 *is the same as in Lemma 2.1.*

Proof. By using the differential formula, we have

$$\begin{aligned}\text{div}G(x) &= \sum_{i=1}^n \frac{\partial G_i(x)}{\partial x_i} = \int_0^1 \sum_{i=1}^n \frac{\partial \phi_\epsilon}{\partial x_i}(x - p_1 + t(p_1 - p_2))(p_{1i} - p_{2i})dt \\ &= \int_0^1 \nabla\phi_\epsilon(x - p_1 + t(p_1 - p_2))(p_1 - p_2)dt \\ &= \int_0^1 \frac{d}{dt}\phi_\epsilon(x - p_1 + t(p_1 - p_2))dt \\ &= \phi_\epsilon(x - p_2) - \phi_\epsilon(x - p_1).\end{aligned}$$

Since $(1 - t)p_1 + tp_2 \in B(p, \rho - \delta)\,(0 \le t \le 1)$ and $f(x) \notin B(p, \rho)$ $(x \in \partial\Omega)$, we obtain

$$||f(x) - ((1 - t)p_1 + tp_2)|| \ge \delta > \epsilon, \quad 0 \le t \le 1,\ x \in \partial\Omega.$$

Thus,

$$\phi_\epsilon(f(x) - p_1 + t(p_1 - p_2)) = 0, \quad t \in [0, 1],\ x \in \partial\Omega.$$

Hence, $G(f(x)) = 0\,(x \in \partial\Omega)$. This is the end.

Lemma 2.3. *Suppose* $f \in C^2(\bar{\Omega}, \mathbb{R}^n)$. *Let* $A_{ij}(x)$ *be the cofactor of the element* $\frac{\partial f_j}{\partial x_i}$ *in the Jacobian determinant* $J_f(x)$. *Then the following* **Hadamard identity** *holds:*

$$\sum_{i=1}^n \frac{\partial}{\partial x_i} A_{ij}(x) = 0, \quad j = 1, 2, \ldots, n.$$

Proof. Without loss of generality, we only consider the case $j = 1$. Let

$$D_i(x) = \begin{vmatrix} \frac{\partial f_2}{\partial x_1} & \cdots & \frac{\partial f_2}{\partial x_{i-1}} & \frac{\partial f_2}{\partial x_{i+1}} & \cdots & \frac{\partial f_2}{\partial x_n} \\ \vdots & & \vdots & \vdots & & \vdots \\ \frac{\partial f_n}{\partial x_1} & \cdots & \frac{\partial f_n}{\partial x_{i-1}} & \frac{\partial f_n}{\partial x_{i+1}} & \cdots & \frac{\partial f_n}{\partial x_n} \end{vmatrix}.$$

Then $A_{i1}(x) = (-1)^{i+1} D_i(x)$. Using the derivative rule of determinant, we have

$$\frac{\partial A_{i1}(x)}{\partial x_i} = (-1)^{i+1} \frac{\partial D_i(x)}{\partial x_i} = (-1)^{i+1} \sum_{k \neq i} E_{ik}(x),$$

where

$$E_{ik}(x) = \begin{vmatrix} \frac{\partial f_2}{\partial x_1} & \cdots & \frac{\partial^2 f_2}{\partial x_i \partial x_k} & \cdots & \frac{\partial f_2}{\partial x_{i-1}} & \frac{\partial f_2}{\partial x_{i+1}} & \cdots & \frac{\partial f_2}{\partial x_n} \\ \vdots & & \vdots & & \vdots & \vdots & & \vdots \\ \frac{\partial f_n}{\partial x_1} & \cdots & \frac{\partial^2 f_n}{\partial x_i \partial x_k} & \cdots & \frac{\partial f_n}{\partial x_{i-1}} & \frac{\partial f_n}{\partial x_{i+1}} & \cdots & \frac{\partial f_n}{\partial x_n} \end{vmatrix},$$

that is, $E_{ik}(x)$ is obtained through the method that the elements $\frac{\partial f_j}{\partial x_k}$ $(j = 2, \ldots, n)$ in k-column for D_i are replaced by $\frac{\partial^2 f_j}{\partial x_i \partial x_k}$ $(j = 2, \ldots, n)$. Thus,

$$\sum_{i=1}^{n} \frac{\partial}{\partial x_i} A_{i1}(x) = \sum_{i=1}^{n} (-1)^{i+1} \sum_{k \neq i} E_{ik}(x) = \sum_{i \neq k, 1 \leq i, k \leq n} (-1)^{i+1} E_{ik}(x). \tag{2.7}$$

Due to the continuity of the second partial derivative of f, we have $\frac{\partial^2 f_j}{\partial x_i \partial x_k} = \frac{\partial^2 f_j}{\partial x_k \partial x_i}$. Comparing $E_{ik}(x)$ and $E_{ki}(x)$, we conclude that if $i > k$,

$$E_{ik}(x) = (-1)^{i-k-1} E_{ki}(x).$$

Then,

$$(-1)^{i+1} E_{ik}(x) = (-1)^{-k} E_{ki}(x) = -(-1)^{k+1} E_{ki}(x).$$

Hence, the rightmost term in (2.7) cancels out each other pairwise. This is the end.

Lemma 2.4. *Suppose that* $A_{ij}(x)$, $G(x) = (G_1(x), \ldots, G_n(x))$ *are as in the two lemmas mentioned above, respectively. Let*

$$F(x) = (F_1(x), \ldots, F_n(x)),$$
$$F_i(x) = \sum_{j=1}^{n} A_{ij}(x) G_j(f(x)), \quad i = 1, 2, \ldots, n.$$

Then

$$\mathrm{div} F(x) = [\phi_\epsilon(f(x) - p_2) - \phi_\epsilon(f(x) - p_1)] J_f(x).$$

Proof. From definition, it follows that

$$\begin{aligned}
\frac{\partial F_i(x)}{\partial x_i} &= \sum_{j=1}^{n} \left(\frac{\partial}{\partial x_i} A_{ij}(x) \right) G_j(f(x)) + \sum_{j=1}^{n} A_{ij}(x) \frac{\partial}{\partial x_i} G_j(f(x)) \\
&= \sum_{j=1}^{n} \left(\frac{\partial}{\partial x_i} A_{ij}(x) \right) G_j(f(x)) \\
&\quad + \sum_{j=1}^{n} A_{ij}(x) \sum_{k=1}^{n} \frac{\partial}{\partial y_k} G_j(f(x)) \frac{\partial f_k}{\partial x_i}.
\end{aligned}$$

Then,

$$\begin{aligned}
\mathrm{div} F(x) &= \sum_{j=1}^{n} \left[\sum_{i=1}^{n} \frac{\partial}{\partial x_i} A_{ij}(x) \right] G_j(f(x)) \\
&\quad + \sum_{j=1}^{n} \sum_{k=1}^{n} \left[\sum_{i=1}^{n} A_{ij}(x) \frac{\partial f_k}{\partial x_i} \right] \frac{\partial}{\partial y_k} G_j(f(x)) \\
&= \sum_{j,k=1}^{n} \delta_{jk} J_f(x) \frac{\partial}{\partial y_k} G_j(f(x)) \\
&= \sum_{j=k=1}^{n} J_f(x) \frac{\partial}{\partial y_k} G_j(f(x)) \\
&= J_f(x) \mathrm{div} G(f(x)) \\
&= [\phi_\epsilon(f(x) - p_2) - \phi_\epsilon(f(x) - p_1)] J_f(x).
\end{aligned}$$

This is the end.

By Stokes Theorem and Lemma 2.2, we arrive at

$$\int_{\Omega} \operatorname{div} F(x)dx = \int_{\partial\Omega} F(x)dx = 0.$$

This proves the rationality of using (2.2) to define topological degree.

2.1.3 Brouwer Degree

The definition of Brouwer degree has been given for the C^2 mappings. In this subsection, we want to define it for the continuous mappings.

Definition 2.1.3. Suppose that $\Omega \subset \mathbb{R}^n$ is a bounded open subset, $f \in C(\bar{\Omega}, \mathbb{R}^n)$, and $p \in \mathbb{R}^n \backslash f(\partial\Omega)$. Let $f_1 \in C^2(\bar{\Omega}, \mathbb{R}^n)$ and

$$\sup_{x\in\bar{\Omega}} ||f(x) - f_1(x)|| < \rho = \operatorname{dist}(p, f(\partial\Omega)).$$

Define the **Brouwer degree** of f in Ω with respect to p by

$$\deg(f, \Omega, p) = \deg(f_1, \Omega, p). \tag{2.8}$$

Because it has been explained how to define Brouwer degrees for the C^2 function, the right-hand side of (2.8) is well defined. To show the rationality of the definition, it should also be shown that if there is another function $f_2 \in C^2(\bar{\Omega}, \mathbb{R}^n)$ such that

$$\sup_{x\in\bar{\Omega}} ||f(x) - f_2(x)|| < \rho = \operatorname{dist}(p, f(\partial\Omega)),$$

then

$$\deg(f_1, \Omega, p) = \deg(f_2, \Omega, p). \tag{2.9}$$

Let's first provide a lemma and then provide a proof for (2.9).

Lemma 2.5. *Let $\Omega \subset \mathbb{R}^n$ be a bounded open subset, $f \in C^2(\bar{\Omega}, \mathbb{R}^n)$, $p \in \mathbb{R}^n$, and $p \notin f(\partial\Omega)$. Then, for every $g \in C^2(\bar{\Omega}, \mathbb{R}^n)$, there exists $\delta = \delta(f, p, g) > 0$ such that, whenever $|t| < \delta$, we have that $p \notin (f + tg)(\partial\Omega)$, and*

$$\deg(f + tg, \Omega, p) = \deg(f, \Omega, p).$$

Proof. Let $M > \max_{x\in\bar{\Omega}} ||g(x)||$. We consider three cases as follows.

Case 1. $f^{-1}(p) = \emptyset$. Let $\delta = \text{dist}(p, f(\bar{\Omega})) \cdot M^{-1}$. If $|t| < \delta$, for $\forall x \in \bar{\Omega}$, we have

$$\begin{aligned}||f(x) + tg(x) - p|| &\geq ||f(x) - p|| - |t| \cdot ||g(x)|| \\ &\geq \text{dist}(p, f(\bar{\Omega})) - |t|M \\ &> \text{dist}(p, f(\bar{\Omega})) - \delta M = 0.\end{aligned}$$

Thus, $(f + tg)^{-1}(p) = \emptyset$.

Case 2. $f^{-1}(p) \neq \emptyset$ and p is a regular value of f. From the Inverse Function Theorem, $f^{-1}(p)$ is a finite set. Let $f^{-1}(p) = \{x_1, \ldots, x_k\}$, $\rho = \text{dist}(p, f(\partial\Omega))$, and $f_t(x) := F(x,t) := f(x) + tg(x)$. Denote $\delta_1 = \rho \cdot M^{-1}$. Then, if $|t| < \delta_1$, $p \notin f_t(\partial\Omega)$.

Since $\det F'_x(x_i, 0) = J_f(x_i)$ and $F(x_i, 0) = p$, we know from the Implicit Function Theorem that there exist an interval $(-\delta_2, \delta_2)$ and disjoint balls $B_{r_i}(x_i)$ such that, whenever $t \in (-\delta_2, \delta_2)$, the equation

$$F(x,t) = p \tag{2.10}$$

admits a unique continuous solution $x_i(t)$ in $B_{r_i}(x_i)$ and $x_i(0) = x_i$. From the continuity of J_{f_t} with respect to (x,t), we can also require that if $|t| < \delta_2$,

$$\text{sign}J_{f_t}(x_i(t)) = \text{sign}J_f(x_i). \tag{2.11}$$

Let $V = \bigcup_{i=1}^k B_{r_i}(x_i)$. Because the equation $f(x) = p$ has no solution in $\bar{\Omega}\backslash\bar{V}$, we have $\text{dist}(f(\bar{\Omega}\backslash\bar{V}), p) = \beta > 0$. Denote $\delta_3 = \frac{\beta}{M}$. When $|t| < \delta_3$, the equation (2.10) has only k solutions $x_1(t), x_2(t), \ldots, x_k(t)$ in Ω. Combined with (2.11), we arrive at

$$\deg(f_t, \Omega, p) = \sum_{i=1}^{k} \text{sign}J_{f_t}(x_i(t)) = \sum_{i=1}^{k} \text{sign}J_f(x_i) = \deg(f, \Omega, p).$$

Case 3. $f^{-1}(p) \neq \emptyset$ and p is a critical value of f. Take a regular value p_1 of f such that $||p_1 - p|| < \frac{1}{3}\rho$, where $\rho = \text{dist}(p, f(\partial\Omega))$.

From the Case 2, there exists a $\delta > 0$ such that, whenever $|t| < \delta$, $\text{dist}(p, f_t(\partial\Omega)) > \frac{1}{3}\rho$,

$$\deg(f_t, \Omega, p_1) = \deg(f, \Omega, p_1),$$

and p_1 is a regular value of f_t. From the Definition 2.1.2, we obtain

$$\deg(f_t, \Omega, p) = \deg(f_t, \Omega, p_1) = \deg(f, \Omega, p_1) = \deg(f, \Omega, p).$$

This is the end.

Lemma 2.6. *Let $\Omega \subset \mathbb{R}^n$ be a bounded open subset, $f \in C(\bar{\Omega}, \mathbb{R}^n)$, and $p \in \mathbb{R}^n \backslash f(\partial\Omega)$. Suppose $f_1, f_2 \in C^2(\bar{\Omega}, \mathbb{R}^n)$ and*

$$\sup_{x\in\bar{\Omega}} ||f_i(x) - f(x)|| < \rho = \text{dist}(p, f(\partial\Omega)), \quad i = 1, 2.$$

Then

$$\deg(f_1, \Omega, p) = \deg(f_2, \Omega, p).$$

Proof. Let $h(x,t) = tf_1(x) + (1-t)f_2(x)$. Then, for any fixed $t \in [0,1]$ and every $x \in \bar{\Omega}$, we have

$$||h(x,t) - f(x)|| \leq t||f_1(x) - f(x)|| + (1-t)||f_2(x) - f(x)|| < \rho.$$

Hence, $p \notin h(\partial\Omega, t)$.

Denote $\phi(t) = \deg(h(\cdot, t), \Omega, p)$. For every $t_0 \in [0,1]$, we know from Lemma 2.5 that there exists a δ-neighborhood of t_0 such that $\phi(t)$ is constant in this δ-neighborhood. Thus, $\phi(t)$ is a continuous function in $[0,1]$. In addition, $\phi(t)$ is an integer valued function. Hence, ϕ is equal to constant in $[0,1]$. Therefore, $\phi(0) = \phi(1)$, i.e.

$$\deg(f_1, \Omega, p) = \deg(f_2, \Omega, p).$$

This is the end.

The above definition of Brouwer degree also provides an algorithm.

Example 2.1. Suppose that A is an $n \times n$ real coefficient nondegenerate matrix and $0 \in \Omega$. Then

$$\deg(A, \Omega, 0) = (-1)^{\beta},$$

where $\beta = \sum_{\lambda_j<0} \beta_j$ and β_j are the algebraic multiplicities of the negative eigenvalues λ_j of A, i.e.,

$$\beta_j = \dim \bigcup_{k=1}^{\infty} \ker(\lambda_j I - A)^k.$$

Proof. Let $\lambda_1, \ldots, \lambda_n$ be all the eigenvalues of A. By definition,

$$\deg(A, \Omega, 0) = \operatorname{sign} \det A = \operatorname{sign} \prod_{j=1}^{n} \lambda_j.$$

Since complex roots appear in pairs and positive eigenvalues have no contribution to $\operatorname{sign}\det A$, hence

$$\operatorname{sign} \prod_{j=1}^{n} \lambda_j = \operatorname{sign} \prod_{\lambda_j<0} \lambda_j = \prod_{\lambda_j<0} (-1)^{\beta_j} = (-1)^{\beta}.$$

This is the end.

2.2 Properties of Brouwer Degree

In this section, we always assume that $\Omega \subset \mathbb{R}^n$ is a bounded open subset.

2.2.1 Fundamental Properties of Brouwer Degree

Theorem 2.2. *Let $\Omega \subset \mathbb{R}^n$ be a bounded open subset, $f \in C(\bar{\Omega}, \mathbb{R}^n)$, and $p \in \mathbb{R}^n \backslash f(\partial\Omega)$. If $g \in C(\bar{\Omega}, \mathbb{R}^n)$ satisfies*

$$\sup_{x\in\bar{\Omega}} ||f(x) - g(x)|| < \rho = \operatorname{dist}(p, f(\partial\Omega)),$$

then

$$\deg(f, \Omega, p) = \deg(g, \Omega, p).$$

Proof. It is easy to see $p \notin g(\partial\Omega)$. Let $\rho_1 = \operatorname{dist}(p, g(\partial\Omega))$. Take $h \in C^2(\bar{\Omega}, \mathbb{R}^n)$ satisfying

$$\sup_{x\in\bar{\Omega}} ||h(x) - g(x)|| < \min\{\rho_1, \rho - ||f - g||\},$$

where $||f - g|| = \sup_{x\in\bar{\Omega}} ||f(x) - g(x)||$. Then,

$$\sup_{x\in\bar{\Omega}} ||f(x) - h(x)|| \leq \sup_{x\in\bar{\Omega}} ||f(x) - g(x)|| + \sup_{x\in\bar{\Omega}} ||g(x) - h(x)|| < \rho.$$

By Definition 2.1.3,

$$\deg(f, \Omega, p) = \deg(h, \Omega, p) = \deg(g, \Omega, p).$$

This is the end.

Theorem 2.3. *Brouwer degree* $\deg(f, \Omega, p)$ *possesses the following properties.*

(1) (**Normality**)

$$\deg(id, \Omega, p) = \begin{cases} 1, & p \in \Omega, \\ 0, & p \notin \bar{\Omega}. \end{cases}$$

(2) (**Additivity**) *Let* Ω_1, Ω_2 *be open subsets of* Ω *such that* $\Omega_1 \cap \Omega_2 = \emptyset$ *and* $p \notin f(\bar{\Omega}\backslash(\Omega_1 \cup \Omega_2))$. *Then*

$$\deg(f, \Omega, p) = \deg(f, \Omega_1, p) + \deg(f, \Omega_2, p).$$

(3) (**Homotopy invariance**) *Let* $H : \bar{\Omega} \times [0, 1] \to \mathbb{R}^n$ *be continuous and* $h_t(x) = H(x, t)$. *Suppose that* $p : [0, 1] \to \mathbb{R}^n$ *is continuous, and* $p(t) \notin h_t(\partial\Omega)$ *for* $t \in [0, 1]$. *Then*

$$\deg(h_t, \Omega, p(t)) \text{ is independent of } t.$$

Remark. The above three properties are the fundamental properties of degree, which can serve as an axiom system for establishing degree theory.

Proof. (1) It is obvious.

(2) Since f is continuous and $\Omega_0 = \bar{\Omega}\backslash(\Omega_1 \cup \Omega_2)$ is a bounded closed subset, we know from $p \notin f(\Omega_0)$ that

$$\inf_{x\in\Omega_0} ||f(x) - p|| =: \rho > 0.$$

Take $f_1 \in C^2(\bar{\Omega}, \mathbb{R}^n)$ satisfying $\sup_{x\in\bar{\Omega}} ||f(x) - f_1(x)|| < \rho$. Then $p \notin f_1(\Omega_0)$. Thus,

$$\inf_{x\in\Omega_0} ||f_1(x) - p|| =: \rho_1 > 0.$$

Noticing $\partial\Omega, \partial\Omega_1, \partial\Omega_2 \subset \Omega_0$, we conclude from Definition 2.1.3 that

$$\deg(f, \Omega, p) = \deg(f_1, \Omega, p), \tag{2.12}$$

$$\deg(f, \Omega_i, p) = \deg(f_1, \Omega_i, p), \quad i = 1, 2. \tag{2.13}$$

Take p_1 such that p_1 is a regular value of f_1 and $||p - p_1|| < \rho_1$. Then $p_1 \notin f_1(\Omega_0)$. From Definition 2.1.2, it follows that

$$\deg(f_1, \Omega, p) = \deg(f_1, \Omega, p_1), \tag{2.14}$$

$$\deg(f_1, \Omega_i, p) = \deg(f_1, \Omega_i, p_1), \quad i = 1, 2. \tag{2.15}$$

Since $f_1^{-1}(p_1) \subset \Omega_1 \cup \Omega_2$, we conclude from Definition 2.1.2 that

$$\begin{aligned}
\deg(f_1, \Omega, p_1) &= \sum_{x\in f_1^{-1}(p_1)} \operatorname{sign} J_{f_1}(x) \\
&= \sum_{x\in f_1^{-1}(p_1)\cap\Omega_1} \operatorname{sign} J_{f_1}(x) + \sum_{x\in f_1^{-1}(p_1)\cap\Omega_2} \operatorname{sign} J_{f_1}(x) \\
&= \deg(f_1, \Omega_1, p_1) + \deg(f_1, \Omega_2, p_1).
\end{aligned}$$

Combined with (2.12)–(2.15), we obtain

$$\deg(f, \Omega, p) = \deg(f, \Omega_1, p) + \deg(f, \Omega_2, p).$$

(3) By the definition of Brouwer degree, we know that if $g(x) = f(x) - p$, then $\deg(f, \Omega, p) = \deg(g, \Omega, 0)$. Thus, we may assume that $p(t) = p$ is independent of t.

For any fixed $t_0 \in [0, 1]$, we obtain from assumption that $\sup_{x\in\bar{\Omega}} ||h_t(x) - h_{t_0}(x)|| \to 0$ as $t \to t_0$. By Theorem 2.2, $\phi(t) = \deg(h_t, \Omega, p)$ is constant in a neighborhood of t_0. Thus, $\phi(t)$ is a continuous function in $[0, 1]$. From the integer value of ϕ and the connectivity of $[0, 1]$, it can be concluded that $\phi(t)$ is independent of t. This is the end.

2.2.2 Properties of Brouwer Degree

Using Theorem 2.3, it is easy to obtain the following theorems.

Theorem 2.4. *The Brouwer degree* $\deg(f, \Omega, p)$ *possesses the following properties.*

(1) (**Excision**) *If* $K \subset \bar{\Omega}$ *is closed and* $p \notin f(K)$, *then*

$$\deg(f, \Omega, p) = \deg(f, \Omega \backslash K, p).$$

(2) (**Kronecker existence**) *If* $p \notin f(\bar{\Omega})$, *then* $\deg(f, \Omega, p) = 0$. *If* $\deg(f, \Omega, p) \neq 0$, *then the equation* $f(x) = p$ *admits at least one solution in* Ω.
(3) (**Connected components**) *When* p *moves within a connected component of* $\mathbb{R}^n \backslash f(\partial\Omega)$, $\deg(f, \Omega, p)$ *does not change (is constant). Particularly, if* p *is in an unbounded component,* $\deg(f, \Omega, p) = 0$.
(4) (**Boundary value**) $\deg(f, \Omega, p)$ *only depends on restriction of* f *on* $\partial\Omega$, *i.e., if* $g \in C(\bar{\Omega}, \mathbb{R}^n)$ *and* $g\big|_{\partial\Omega} = f\big|_{\partial\Omega}$, *then*

$$\deg(f, \Omega, p) = \deg(g, \Omega, p).$$

Theorem 2.5 (Poincaré–Bohl). *Let* $f, g \in C(\bar{\Omega}, \mathbb{R}^n)$. *For any* $x \in \partial\Omega$, *suppose that* p *is not on the line segment connected by* $f(x)$ *and* $g(x)$. *Then*

$$\deg(f, \Omega, p) = \deg(g, \Omega, p).$$

Proof. Let

$$h_t(x) = H(x, t) = (1 - t)f(x) + tg(x), \quad 0 \leq t \leq 1.$$

Then, whenever $0 \leq t \leq 1$, $p \notin h_t(\partial\Omega)$. The conclusion is from the homotopy invariance. This is the end.

Theorem 2.6 (Acute-angle Principle). *Let* $0 \in \Omega \subset \mathbb{R}^n$. *Suppose that whenever* $x \in \partial\Omega$, *the inner product* $\langle x, f(x) \rangle > 0$. *Then* $\deg(f, \Omega, 0) = 1$.

Proof. It is easy to see $f(x) \neq 0$ on $\partial\Omega$. Let

$$h_t(x) = (1 - t)x + tf(x), \quad \forall x \in \bar{\Omega}, \ 0 \leq t \leq 1.$$

Thus

$$||h_t(x)||^2 = (1-t)^2||x||^2 + 2t(1-t)\langle f(x), x\rangle + t^2||f(x)||^2.$$

From the assumption, it follows that, whenever $x \in \partial\Omega$ and $0 \le t \le 1$, $||h_t(x)||^2 > 0$. Thus, $0 \notin h_t(\partial\Omega)$, $0 \le t \le 1$. By the homotopy invariance and normality, we arrive at

$$\deg(f, \Omega, 0) = \deg(h_1, \Omega, 0) = \deg(h_0, \Omega, 0) = \deg(id, \Omega, 0) = 1.$$

This is the end.

Theorem 2.7 (Lack of Direction). *Suppose that* $0 \notin f(\partial\Omega)$ *and there exists a fixed* $x_0 \in \mathbb{R}^n \backslash \{0\}$ *such that, whenever* $x \in \partial\Omega$ *and* $\lambda > 0$, $f(x) \neq \lambda x_0$. *Then* $\deg(f, \Omega, 0) = 0$.

Proof. We prove it by arguing a contradiction. Suppose $\deg(f, \Omega, 0) \neq 0$. Take τ_0 such that $\tau_0 > \sup_{x \in \bar{\Omega}} ||f(x)|| \cdot ||x_0||^{-1}$. Let $H(t, x) = f(x) - t\tau_0 x_0$. Then, whenever $0 \le t \le 1$ and $x \in \partial\Omega$, $H(t, x) \neq 0$. From the homotopy invariance,

$$\deg(f, \Omega, 0) = \deg(f - \tau_0 x_0, \Omega, 0) \neq 0.$$

By Kronecker existence, there exists $x_1 \in \Omega$ such that $f(x_1) - \tau_0 x_0 = 0$, which is in contradiction with the choice of τ_0. This is the end.

Theorem 2.8 (Dimensionality Reduction). *Suppose that* f *maps* $\bar{\Omega}$ *into a lower dimensional subspace* $\mathbb{R}^m$ $(m < n)$. *Then, for every* $p \in \mathbb{R}^n \backslash f(\partial\Omega)$, $\deg(f, \Omega, p) = 0$.

Proof. If $p \notin \mathbb{R}^m$, then $p \notin f(\bar{\Omega})$. Thus, the Kronecker existence implies $\deg(f, \Omega, p) = 0$.

If $p \in \mathbb{R}^m$, take $x_0 \in \mathbb{R}^n \backslash \mathbb{R}^m$ and $x_0 \neq 0$. Then,

$$x \in \partial\Omega,\ \lambda > 0 \Longrightarrow f(x) - p \neq \lambda x_0.$$

Thus,

$$\deg(f, \Omega, p) = \deg(f - p, \Omega, 0) = 0.$$

This is the end.

2.2.3 The Reduction Theorem and Product Formula

Theorem 2.9 (Reduction Theorem). *Let $m < n$, $\mathbb{R}^m$ be a subspace of $\mathbb{R}^n$, and $\Omega \subset \mathbb{R}^n$ be a bounded open subset. Suppose that $F \in C(\bar{\Omega}, \mathbb{R}^m)$, $f(x) = x - F(x)$, and $p \in \mathbb{R}^m \backslash f(\partial\Omega)$. Then*

$$\deg(f, \Omega, p) = \deg(g, \Omega \cap \mathbb{R}^m, p), \tag{2.16}$$

where $g = f\big|_{\Omega\cap\mathbb{R}^m}$.

Proof. Suppose $\Omega \cap \mathbb{R}^m = \emptyset$. Then, the right-hand side of (2.16) is 0. Because $p \in \mathbb{R}^m$ implies $p \notin f(\bar{\Omega})$, thus the left-hand side of (2.16) is 0.

Suppose $\Omega \cap \mathbb{R}^m \neq \emptyset$. Since $p \in \mathbb{R}^m$ and $F(\bar{\Omega}) \subset \mathbb{R}^m$, every solution x of the equation $f(x) = p$ satisfies $x = p + F(x) \in \mathbb{R}^m$. Thus, $f^{-1}(p) = g^{-1}(p)$.

Assume that $f \in C^2(\bar{\Omega}, \mathbb{R}^n)$ and p is a regular value of f. Let's assume that

$$F(x) = (F_1(x), \ldots, F_m(x), 0, \ldots, 0).$$

Then, we can denote

$$f'(x) = \begin{pmatrix} g'(x) & * \\ 0 & id \end{pmatrix}.$$

Thus, p is also a regular value of g and

$$\text{sign} J_f(x) = \text{sign} J_g(x),$$

where $x \in f^{-1}(p)$. By Definition 2.1.2,

$$\begin{aligned} \deg(f, \Omega, p) &= \sum_{x\in f^{-1}(p)} \text{sign} J_f(x) = \sum_{x\in g^{-1}(p)} \text{sign} J_g(x) \\ &= \deg(g, \Omega \cap \mathbb{R}^m, p). \end{aligned} \tag{2.17}$$

Assume that p is not a regular value of f. Take $p_1 \in \mathbb{R}^m$ such that p_1 is a regular value of g and satisfies $||p - p_1|| < \rho = \text{dist}(p, f(\partial\Omega)) \leq \text{dist}(p, g(\partial(\Omega \cap \mathbb{R}^m)))$. Thus, p_1 is also a regular value of f. By Definition 2.1.2 and (2.17),

$$\begin{aligned} \deg(f, \Omega, p) &= \deg(f, \Omega, p_1) = \deg(g, \Omega \cap \mathbb{R}^m, p_1) \\ &= \deg(g, \Omega \cap \mathbb{R}^m, p). \end{aligned} \tag{2.18}$$

Assume that f is a continuous function. Take $F_1 \in C^2(\bar{\Omega}, \mathbb{R}^m)$ such that

$$\sup_{x\in\bar{\Omega}} ||F(x) - F_1(x)|| < \rho = \text{dist}(p, f(\partial\Omega)).$$

Let $f_1(x) = x - F_1(x)$. By Definition 2.1.3, (2.17), and (2.18),

$$\deg(f, \Omega, p) = \deg(f_1, \Omega, p) = \deg(g_1, \Omega \cap \mathbb{R}^m, p) = \deg(g, \Omega \cap \mathbb{R}^m, p),$$

where g_1 is the restriction f_1 on $\Omega \cap \mathbb{R}^m$. This is the end.

Theorem 2.10 (Product Formula). *Let $\Omega, D \subset \mathbb{R}^n$ be bounded open subsets, $f \in C(\bar{\Omega}, \mathbb{R}^n)$, $f(\bar{\Omega}) \subset D$, $g \in C(\bar{D}, \mathbb{R}^n)$, and D_α be the bounded connected components of $D\backslash f(\partial\Omega)$. Suppose that $p \notin (g \circ f)(\partial\Omega) \cup g(\partial D)$. Then*

$$\deg(g \circ f, \Omega, p) = \sum_\alpha \deg(g, D_\alpha, p) \deg(f, \Omega, D_\alpha),$$

where $\deg(f, \Omega, D_\alpha) = \deg(f, \Omega, q)$, $q \in D_\alpha$, *is independent of the choice of q from the properties of connected components.*

The proof of this theorem is relatively long, and interested readers can refer to other books [3, 10, 13, 28].

2.2.4 Axiomatization of Degree Theory

In this subsection, we want to prove that Definition 2.1.2 of degree can be derived from the three fundamental properties of Brouwer degree (Theorem 2.3). Therefore, the three fundamental properties of Theorem 2.3 can serve as the definition of Brouwer degree. All the other properties are their consequences. It also indicates that Brouwer degree is a unique existence.

Let's first discuss the homotopy relationship between nondegenerate real matrices. Let $GL(n, \mathbb{R})$ be the set of all $n \times n$ nondegenerate real matrices. Denote by I_n the $n \times n$ identity matrix. Let the matrices $\tilde{I}_1 = (-1)$ and

$$\tilde{I}_n = \begin{pmatrix} I_{n-1} & \\ & -1 \end{pmatrix},$$

where $n > 1$.

Definition 2.2.1. Two matrices A, $B \in GL(n, \mathbb{R})$ are called to be **homotopy**, if there is a continuous matrix function $H : [0,1] \to GL(n, \mathbb{R})$ such that $H(0) = A, H(1) = B$. At this time, we say that H is a **homotopy** of connecting matrices A and B, denoted by $A \overset{H}{\simeq} B$. We will omit H and write $A \simeq B$ for brevity.

For every $t \in [0,1]$, the homotopy $H(t)$ is a real $n \times n$ matrix, and $\det H(t) \neq 0$. It is easy to show that the homotopy relation $\simeq$ of matrices is an equivalence relation on $GL(n, \mathbb{R})$. In the sequel, we first establish some simple propositions on matrix homotopy relationships.

Lemma 2.7. *The following three statements hold:*

(i) *Let $A, C \in GL(m, \mathbb{R})$ and $B, D \in GL(n, \mathbb{R})$. Suppose $A \simeq C$ and $B \simeq D$. Then*

$$\begin{pmatrix} A & \\ & B \end{pmatrix} \simeq \begin{pmatrix} C & \\ & D \end{pmatrix}.$$

(ii) *Suppose $A, B, P, Q \in GL(n, \mathbb{R})$. If $A \overset{H}{\simeq} B$, then $PAQ \overset{PHQ}{\simeq} PBQ$.*

(iii) $I_{2m} \simeq -I_{2m}$.

Proof. (i) and (ii) are clear. Now, we prove (iii). We define a matrix function as follows.

$$H : [0,1] \to GL(2m, \mathbb{R}),\ t \mapsto \begin{pmatrix} \cos \pi t \cdot I_m & \sin \pi t \cdot I_m \\ -\sin \pi t \cdot I_m & \cos \pi t \cdot I_m \end{pmatrix}.$$

A direct computation provides that, for every $t \in [0,1]$,

$$[H(t)]^{-1} = \begin{pmatrix} \cos \pi t \cdot I_m & -\sin \pi t \cdot I_m \\ \sin \pi t \cdot I_m & \cos \pi t \cdot I_m \end{pmatrix}.$$

Thus, H is a homotopy connecting I_{2m} and $-I_{2m}$. This is the end.

Theorem 2.11. *Matrices $A, B \in GL(n, \mathbb{R})$ are homotopy, if and only if their determinants have the same sign, that is,*

$$A \simeq B \Leftrightarrow \det AB > 0.$$

Proof. Clearly, the homotopy matrices have the same determinant sign. We will prove sufficiency, that means to prove the following conclusions: for every matrix $A \in GL(n,\mathbb{R})$,

$$\det A > 0 \Rightarrow A \simeq I_n, \quad \det A < 0 \Rightarrow A \simeq \tilde{I}_n.$$

Our approach is to use homotopy to simplify the matrix A to I_n or $\tilde{I}_n$ step by step. We investigate the general case where A has positive, complex, and negative eigenvalues, simultaneously. From the theory of Jordan canonical form of a matrix, there exists a matrix $P \in GL(n,\mathbb{R})$ such that $A = PJP^{-1}$, where the matrix J admits the following form

$$J = \begin{pmatrix} J_+ & & & \\ & \text{Re } C & \text{Im } C & \\ & -\text{Im } C & \text{Re } C & \\ & & & J_- \end{pmatrix} \in GL(n,\mathbb{R}),$$

J_+, J_- are triangular real $n_+ \times n_+, n_- \times n_-$ matrices composed of Jordan blocks corresponding to the positive and negative eigenvalues of A, respectively, $\begin{pmatrix} C & \\ & \overline{C} \end{pmatrix}$ is a triangular complex $2n_c \times 2n_c$ matrix composed of Jordan blocks corresponding to the complex eigenvalues of A, where $n_+ + 2n_c + n_- = n$.

Let the matrix $D = \begin{pmatrix} I_{n_+} & & \\ & I_{2n_c} & \\ & & -I_{n_-} \end{pmatrix}$. Define a matrix function

$$\begin{aligned} h(t) &= (1-t)I_{2n_c} + t\begin{pmatrix} \text{Re } C & \text{Im } C \\ -\text{Im } C & \text{Re } C \end{pmatrix} \\ &= \begin{pmatrix} \text{Re }[(1-t)I_{n_c} + tC] & \text{Im }[(1-t)I_{n_c} + tC] \\ -\text{Im }[(1-t)I_{n_c} + tC] & \text{Re }[(1-t)I_{n_c} + tC] \end{pmatrix}. \end{aligned}$$

For each $t \in [0,1]$, the matrices $h(t)$ and $\begin{pmatrix} (1-t)I_{n_c}+tC & \\ & (1-t)I_{n_c}+t\overline{C} \end{pmatrix}$ are complex similarity, so that $h(t)$ is invertible. Thus, h is a homotopy connecting matrices I_{2n_c} and $\begin{pmatrix} \text{Re } C & \text{Im } C \\ -\text{Im } C & \text{Re } C \end{pmatrix}$. Hence, the matrix function $H(t) := (1-t)D + tJ$ is a homotopy connecting matrices D and J. Therefore, $A \simeq PDP^{-1}$.

In the sequel, we will consider the signs of $\det A$ and $\det P$, odd or even of n to discuss several cases as follows.

(i) Assume $\det A > 0$. Then n_- is even. Lemma 2.7 implies $I_{n_-} \simeq -I_{n_-}$ and $D \simeq I_n$. Hence, $A \simeq PI_nP^{-1} = I_n$.

(ii) Assume $\det A < 0$. Then n_- is odd. This implies $D \simeq \tilde{I}_n$. We will consider the sign of $\det P$ again.

(ii.a) Assume $\det P > 0$. The above conclusion (i) implies $P \simeq P^{-1} \simeq I_n$. Thus, $A \simeq I_n \cdot \tilde{I}_n \cdot I_n = \tilde{I}_n$.

(ii.b) Assume $\det P < 0$. We will consider the odd or even of n again.

(ii.b.1) Assume that n is odd. Then, $\det(-P) > 0$. Thus, $A \simeq (-P)D(-P^{-1}) \simeq I_n \cdot \tilde{I}_n \cdot I_n = \tilde{I}_n$.

(ii.b.2) Assume that n is even. Denote

$$Q = \begin{cases} \begin{pmatrix} 0 & 1 \\ 1 & 0 \end{pmatrix}, & n = 2; \\ \begin{pmatrix} 0 & \cdots & 1 \\ \vdots & I_{n-2} & \vdots \\ 1 & \cdots & 0 \end{pmatrix}, & n > 2. \end{cases}$$

Then $\det Q < 0$ and $Q^{-1} = Q$. A computation implies

$$Q\tilde{I}_nQ = \begin{cases} \begin{pmatrix} -1 & 0 \\ 0 & 1 \end{pmatrix}, & n = 2; \\ \begin{pmatrix} -1 & \cdots & 0 \\ \vdots & I_{n-2} & \vdots \\ 0 & \cdots & 1 \end{pmatrix}, & n > 2. \end{cases}$$

Clearly,

$$H_1 : [0,1] \to GL(2,\mathbb{R}), \quad t \mapsto \begin{pmatrix} -\cos \pi t & \sin \pi t \\ \sin \pi t & \cos \pi t \end{pmatrix}$$

is a homotopy connecting the matrices $\left(\begin{smallmatrix} -1 & 0 \\ 0 & 1 \end{smallmatrix}\right)$ and $\left(\begin{smallmatrix} 1 & 0 \\ 0 & -1 \end{smallmatrix}\right)$. Hence, $Q\tilde{I}_nQ \simeq \tilde{I}_n$. Since $\det PQ > 0$, it follows from the conclusion (i) that $PQ \simeq I_n \simeq QP^{-1}$. Therefore,

$$A \simeq P\tilde{I}_nP^{-1} = (PQ)(Q\tilde{I}_nQ)(QP^{-1}) \simeq I_n \cdot \tilde{I}_n \cdot I_n = \tilde{I}_n.$$

This is the end.

In the sequel, we will compute the Brouwer degree for a special matrix only by using the fundamental properties of Brouwer degree in Theorem 2.3.

Lemma 2.8.

$$\deg(\tilde{I}_n, B_r(0), 0) = -1,$$

where $B_r(0) \subset \mathbb{R}^n$ is a ball with radius r centered at 0.

Proof. Denote

$$Q_1 = \begin{pmatrix} 1 & & & \\ & \ddots & & \\ & & 1 & \\ & & & 0 \end{pmatrix}, \quad Q_2 = \begin{pmatrix} 0 & & & \\ & \ddots & & \\ & & 0 & \\ & & & 1 \end{pmatrix}$$

and

$$D = \{\lambda e_n;\ \lambda \in (-2, 2)\}, \quad D_1 = \{\lambda e_n;\ \lambda \in (-2, 0)\},$$
$$D_2 = \{\lambda e_n;\ \lambda \in (0, 2)\},$$

where $e_n = (0, \ldots, 0, 1)^T$. Q_1, Q_2 are bounded linear operators from $\mathbb{R}^n$ to $\mathbb{R}^n$. In the sequel, we compute

$$\deg(Q_1 - Q_2, \tilde{B}_r(0) + D, 0),$$

where $\tilde{B}_r(0) = \{x \in \mathbb{R}^n;\ ||x|| < r, x_n = 0\}$, $\tilde{B}_r(0) + D = \{x + y;\ x \in \tilde{B}_r(0), y \in D\}$.

Denote $f(\lambda e_n) := (|\lambda| - 1)e_n$, $h_t(\lambda e_n) := t(|\lambda| - 2)e_n + e_n$, and $H_t := Q_1 + h_t \circ Q_2$. It is easy to see that whenever $0 \le t \le 1$ and $x \in \partial(\tilde{B}_r(0) + D)$, $H_t(x) \ne 0$. By the homotopy invariance, the Brouwer degree

$$\deg(Q_1 + h_t \circ Q_2, \tilde{B}_r(0) + D, 0)$$

is independent of t. When $t = 0$, it is equal to $\deg(Q_1 + e_n, \tilde{B}_r(0) + D, 0) = 0$; when $t = 1$, it is equal to $\deg(Q_1 + f \circ Q_2, \tilde{B}_r(0) + D, 0)$.

Clearly, $\tilde{B}_r(0) + D_1$ and $\tilde{B}_r(0) + D_2$ are disjoint open subsets of $\tilde{B}_r(0) + D$. By the additivity,

$$\begin{aligned}0 &= \deg(Q_1 + f \circ Q_2, \tilde{B}_r(0) + D, 0) \\ &= \deg(Q_1 + f \circ Q_2, \tilde{B}_r(0) + D_1, 0) \\ &\quad + \deg(Q_1 + f \circ Q_2, \tilde{B}_r(0) + D_2, 0).\end{aligned}$$

Clearly, $f\big|_{D_1}(\lambda e_n) = -(\lambda + 1)e_n$ has a unique zero point $-e_n \in D_1 \subset D$. Thus,

$$\begin{aligned}\deg(Q_1 + f \circ Q_2, \tilde{B}_r(0) + D_1, 0) &= \deg(Q_1 - Q_2 - e_n, \tilde{B}_r(0) + D, 0) \\ &= \deg(Q_1 - Q_2, \tilde{B}_r(0) + D, 0).\end{aligned}$$

The latter equality above holds because there is $Q_1x - Q_2x - te_n \neq 0$ on $\partial(\tilde{B}_r(0) + D) \times [0, 1]$ and homotopy invariance. Similarly, we can prove

$$\deg(Q_1 + f \circ Q_2, \tilde{B}_r(0) + D_2, 0) = \deg(Q_1 + Q_2, \tilde{B}_r(0) + D, 0).$$

Since $Q_1x - Q_2x = 0$ has a unique zero point $x = 0$, then

$$\deg(Q_1 - Q_2, B_r(0), 0) = -1.$$

This is the end.

With the above preparation, we can now prove that the fundamental properties of Brouwer degree can imply Definition 2.1.2. In the sequel, we always assume that $\Omega \subset \mathbb{R}^n$ is a bounded open subset, $f : \bar{\Omega} \to \mathbb{R}^n$ is C^2 continuous, $p \in \mathbb{R}^n$ is a regular value of f, and $p \notin f(\partial\Omega)$. According to the Implicit Function Theorem, $f^{-1}(p) = \{x^1, \ldots, x^k\}$ is composed of a finite number of points. Thus, for each x^i, $f'(x^i) : \mathbb{R}^n \to \mathbb{R}^n$ is an invertible linear mapping.

Lemma 2.9. *Suppose that* deg *satisfies the three fundamental properties in Theorem* 2.3. *Then*

$$\deg(f, \Omega, p) = \sum_{i=1}^{k} \deg(f'(x^i), B_r(0), 0),$$

where $B_r(0)$ *is a ball with radius* r *centered at* 0.

Proof. By the second property (Additivity) in Theorem 2.3, we can choose a small neighborhood U_i of x^i such that these small neighborhoods are disjoint with each other and

$$\deg(f, \Omega, p) = \sum_{i=1}^{k} \deg(f, U_i, p).$$

In the sequel, we will compute $\deg(f, U_i, p)$ for each i. By the definition of F-derivative,

$$f(x) = p + f'(x^i)(x - x^i) + o(||x - x^i||), \quad \text{as } ||x - x^i|| \to 0.$$

Since $\det f'(x^i) \neq 0$, there exists a constant $c > 0$ such that $||f'(x^i)z|| \geq c||z||$, $\forall z \in \mathbb{R}^n$. Denote

$$p(t) = tp, \quad h(x,t) = tf(x) + (1-t)f'(x^i)(x - x^i).$$

For any $t \in [0,1]$ and $||x - x^i|| = \delta$ with small enough δ, we have

$$\begin{aligned} ||h(x,t) - p(t)|| &= ||f'(x^i)(x - x^i) + t \cdot o(||x - x^i||)|| \\ &\geq c||x - x^i|| - o(||x - x^i||) > 0. \end{aligned}$$

From the third property (Homotopy invariance) in Theorem 2.3, it follows that

$$\deg(f, B_\delta(x^i), p) = \deg(f'(x^i) - f'(x^i)x^i, B_\delta(x^i), 0).$$

Since $f(x) \neq p$ in $\bar{U}_i \backslash B_\delta(x^i)$, by the Additivity, we arrive at

$$\deg(f, U_i, p) = \deg(f, B_\delta(x^i), p).$$

Then,

$$\deg(f, U_i, p) = \deg(f'(x^i) - f'(x^i)x^i, B_\delta(x^i), 0).$$

Since x^i is a unique solution of $f'(x^i)x - f'(x^i)x^i = 0$, by the Additivity, for any $B_r(0) \supset B_\delta(x^i)$, we have

$$\deg(f'(x^i) - f'(x^i)x^i, B_\delta(x^i), 0) = \deg(f'(x^i) - f'(x^i)x^i, B_r(0), 0).$$

Because $f'(x^i)(x - tx^i) \neq 0$ on $(x,t) \in \partial B_r(0) \times [0,1]$, by the Homotopy invariance, we obtain

$$\deg(f, U_i, p) = \deg(f'(x^i), B_r(0), 0).$$

By the Additivity again, we know that $r > 0$ can be arbitrary. This is the end.

Theorem 2.12. *Suppose that* deg *satisfies the three fundamental properties in Theorem* 2.3. *Then*

$$\deg(f, \Omega, p) = \sum_{i=1}^{k} \operatorname{sign} \det f'(x^i).$$

Proof. Since $f'(x^i)$ is an invertible linear mapping from $\mathbb{R}^n$ to $\mathbb{R}^n$, it suffices to prove that for each $n \times n$ invertible matrix A, we have

$$\deg(A, B_r(0), 0) = \operatorname{sign} \det A.$$

The previous discussion has shown that if $\det A > 0$, then A is homotopy with I_n, and therefore the conclusion is correct. If $\det A < 0$, then A is homotopy with $\tilde{I}_n$, and therefore the conclusion also is correct. This is the end.

2.2.5 Remark

The definition and properties of Brouwer degree were discussed in $\mathbb{R}^n$ earlier. Now, for finite dimensional normed linear spaces, we can also define Brouwer degree as follows.

Suppose that E_n is an n-dimensional normed linear space, $\Omega \subset E_n$ is an open bounded subset, $f : \bar{\Omega} \to E_n$ is continuous, and $p \in E_n \backslash f(\partial\Omega)$. Let $e_1, e_2, \cdots, e_n$ be a base of E_n. Then, every $x \in E_n$ can be written uniquely as

$$x = \sum_{i=1}^{n} \alpha_i e_i, \quad \text{where } \alpha_i \ (i = 1, 2, \ldots, n) \text{ are real numbers.}$$

Define a mapping $h : E_n \to \mathbb{R}^n$ by $h(x) = (\alpha_1, \ldots, \alpha_n)^T \in \mathbb{R}^n$. Thus, $h : E_n \to \mathbb{R}^n$ is a linear homeomorphism. Hence, $h(\Omega) \subset \mathbb{R}^n$ is a bounded open subset, and $\overline{h(\Omega)} = h(\bar{\Omega})$. Denote $F = hfh^{-1}$.

Clearly, $F : \overline{h(\Omega)} \to \mathbb{R}^n$ is continuous, and $h(p) \in \mathbb{R}^n \backslash F(\partial h(\Omega))$. Therefore, $\deg(F, h(\Omega), h(p))$ is well defined. We denote

$$\deg(f, \Omega, p) = \deg(F, h(\Omega), h(p)).$$

We can prove that $\deg(F, h(\Omega), h(p))$ is independent of the selection of bases.

Suppose that E_n admits another base $\tilde{e}_1, \ldots, \tilde{e}_n$, and each $x \in E_n$ can be written uniquely by

$$x = \sum_{i=1}^{n} \tilde{\alpha}_i \tilde{e}_i, \quad \text{where } \tilde{\alpha}_i \ (i = 1, 2, \ldots, n) \text{ are real numbers.}$$

Define a mapping $\tilde{h} : E_n \to \mathbb{R}^n$ by $\tilde{h}(x) = (\tilde{\alpha}_1, \ldots, \tilde{\alpha}_n)^T \in \mathbb{R}^n$. Then, $\tilde{h} : E_n \to \mathbb{R}^n$ is linear homeomorphism. Denote $\tilde{F} = \tilde{h} f \tilde{h}^{-1}$. Clearly, $\tilde{F} : \overline{\tilde{h}(\Omega)} \to \mathbb{R}^n$ is continuous, and $\tilde{h}(p) \in \mathbb{R}^n \backslash \tilde{F}(\partial \tilde{h}(\Omega))$. Hence, $\deg(\tilde{F}, \tilde{h}(\Omega), \tilde{h}(p))$ is well defined. We need to prove

$$\deg(F, h(\Omega), h(p)) = \deg(\tilde{F}, \tilde{h}(\Omega), \tilde{h}(p)).$$

Assume $(e_1, \ldots, e_n) = (\tilde{e}_1, \ldots, \tilde{e}_n)T$. Thus, $\tilde{\alpha} = T\alpha$, where $\tilde{\alpha} = (\tilde{\alpha}_1, \ldots, \tilde{\alpha}_n)^T \in \mathbb{R}^n$, $\alpha = (\alpha_1, \ldots, \alpha_n)^T \in \mathbb{R}^n$. Then, $\tilde{h}(x) = \tilde{\alpha} = T\alpha = Th(x)$. Therefore,

$$\tilde{h} = Th, \quad \tilde{h}(\bar{\Omega}) = Th(\bar{\Omega}), \quad \tilde{F} = TFT^{-1}.$$

When $\tilde{\alpha} = T\alpha$, we have

$$F(\alpha) = h(p) \Longleftrightarrow \tilde{F}(\tilde{\alpha}) = \tilde{h}(p).$$

According to the definition, we only need to provide a proof for C^2-functions. Assume that F is a C^2-function. Then

$$J_{\tilde{F}}(\tilde{\alpha}) = (\det T) \cdot J_F(T^{-1}\tilde{\alpha}) \cdot (\det T^{-1}) = J_F(\alpha).$$

Hence, when $h(p)$ is a regular value of F, $\tilde{h}(p)$ is a regular value of $\tilde{F}$. By the definition,

$$\deg(F, h(\Omega), h(p)) = \deg(\tilde{F}, \tilde{h}(\Omega), \tilde{h}(p)).$$

If $h(p)$ is a critical value of F, we can take a regular value that is sufficiently close to $h(p)$ and use the definition again.

Remark. So far we have always assumed that the open set $\Omega \subset \mathbb{R}^n$ is bounded. Now, suppose that $\Omega \subset \mathbb{R}^n$ is open but possible unbounded. If we assume that $f \in C(\bar{\Omega}, \mathbb{R}^n)$, $p \notin f(\partial\Omega)$, and $\sup_{\bar{\Omega}} \|x - f(x)\| < \infty$, then $f^{-1}(p)$ will be compact. We can prove that for any bounded open subset $\Omega_0 \supset f^{-1}(p)$, $\deg(f, \Omega \cap \Omega_0, p)$ coincide each other. So we define the degree in unbounded subset by

$$\deg(f, \Omega, p) = \deg(f, \Omega \cap \Omega_0, p),$$

where $\Omega_0 \supset f^{-1}(p)$ is any bounded open subset.

2.3 Brouwer Fixed Point Theorem and Borsuk Theorem

Lemma 2.10. *Suppose that D is a closed convex subset in the Banach space X, $x_0 \in \overset{\circ}{D}$, i.e. x_0 is an interior point of D, and $x_1 \in D$. Then, for every $\lambda \in (0,1)$, $\lambda x_0 + (1-\lambda)x_1 \in \overset{\circ}{D}$. Particularly, if 0 is the interior point of D, then*

$$\lambda x \in \overset{\circ}{D}, \quad \forall x \in D, \lambda \in [0,1).$$

Proof. Assume a ball $B(x_0, r) \subset D$. Take $\epsilon \leq r\lambda$. Then $B(\lambda x_0 + (1-\lambda)x_1, \epsilon) \subset D$. In fact, for every $x \in B(\lambda x_0 + (1-\lambda)x_1, \epsilon)$, we have

$$\begin{aligned} x &= x - [\lambda x_0 + (1-\lambda)x_1] + \lambda x_0 + (1-\lambda)x_1 \\ &= \lambda \left[\frac{x - (\lambda x_0 + (1-\lambda)x_1)}{\lambda} + x_0 \right] + (1-\lambda)x_1 \in D. \end{aligned}$$

This is the end.

Theorem 2.13 (Brouwer Fixed Point Theorem). *Let $\Omega \subset \mathbb{R}^n$ be a bounded closed convex subset, and $\partial\Omega$ be its boundary. Suppose that $f \in C(\Omega, \mathbb{R}^n)$ and $f(\partial\Omega) \subset \Omega$. Then f possesses a fixed point in Ω.*

Proof. Let Ω have an interior point, otherwise, we only need to consider it within the subspace spanned by Ω. It is also possible to assume that 0 is the interior point of Ω, otherwise, take an interior point x_0 of Ω and consider $\Omega - x_0$ and $f(x_0 + x) - x_0$.

If f has a fixed point on $\partial\Omega$, then the theorem naturally holds.

Now we assume that f has no fixed point on $\partial\Omega$. We make the following homotopy

$$h(t, x) = x - tf(x), \quad t \in [0, 1], \ x \in \Omega.$$

At this time, $h(t, x)$ is continuous; if $x \in \partial\Omega$ and $t = 1$, $h(1, x) = x - f(x) \neq 0$; if $x \in \partial\Omega$ and $0 \leq t < 1$, $tf(x) \in \overset{\circ}{\Omega}$ from the assumption, i.e. $tf(x)$ is in the interior of Ω, thus $h(t, x) \neq 0$. Hence, whenever $t \in [0, 1]$, $0 \notin h(t, \partial\Omega)$. By the homotopy invariance and normality, we arrive at

$$\deg(id - f, \overset{\circ}{\Omega}, 0) = \deg(id, \overset{\circ}{\Omega}, 0) = 1.$$

By the Kronecker existence theorem, $id - f$ has a zero point in the interior of Ω, that is, f has a fixed point in the interior of Ω. This is the end.

As an application of Brouwer fixed point theorem, we can prove the following theorem.

Example 2.2 (Perron–Frobenius Theorem). Let $A = (a_{ij})$ be an $n \times n$ matrix such that $a_{ij} \geq 0$, $\forall i, j$. Then there exist $\lambda \geq 0$ and $x \neq 0$ such that $Ax = \lambda x$ and $x_i \geq 0, \forall i$.

Proof. Let

$$D = \left\{ x \in \mathbb{R}^n;\ x_i \geq 0,\ \forall i,\ \sum_{i=1}^{n} x_i = 1 \right\}.$$

If $Ax = 0$ for some $x \in D$, then we take $\lambda = 0$ and confirm the conclusion.

If $Ax \neq 0$ for any $x \in D$, then there exists a positive number $\alpha > 0$ such that

$$\sum_{i=1}^{n} (Ax)_i \geq \alpha, \quad \forall x \in D.$$

Therefore, the mapping

$$f : D \to \mathbb{R}^n, \quad x \mapsto \frac{Ax}{\sum_{i=1}^{n} (Ax)_i}$$

is continuous in D and $f(D) \subset D$. By Brouwer fixed point theorem, f has a fixed point x_0 in D, that is, there exists $x_0 \in D$ such that $Ax_0 = \lambda x_0$, where $\lambda = \sum_{i=1}^{n}(Ax_0)_i$. This is the end.

Example 2.3 (Fundamental Theorem of Algebra). Let $P(z) = a_n z^n + \cdots + a_1 z + a_0$ be a degree n complex polynomial. Then there exists at least one zero point z_0, i.e. $P(z_0) = 0$.

Proof. Without loss of generalization, assume $a_n = 1$. Let

$$M = 2 + |a_0| + |a_1| + \cdots + |a_{n-1}|,$$

and

$$K = \{z \in C;\ |z| \leq M\},$$

where $z = re^{i\theta}$, $0 \leq \theta \leq 2\pi$. We can assume $a_0 \neq 0$, otherwise, 0 will satisfy $P(0) = 0$. Thus, there exists a small enough $\epsilon > 0$ such that

$$|P(z)| > \frac{|a_0|}{2}, \quad |z| \leq \epsilon < 1.$$

By using the continuous extension theorem of continuous mappings, there exists a continuous function $P_\epsilon(z, \bar{z})$ defined on $|z| < \frac{\epsilon}{2}$ such that

$$\hat{P}(z, \bar{z}) = \begin{cases} P_\epsilon(z, \bar{z}), & |z| < \frac{\epsilon}{2}, \\ e^{i(1-n)\theta} P(z), & \frac{\epsilon}{2} \leq |z| \leq 1,\ z = re^{i\theta} \end{cases}$$

is continuous in $|z| \leq 1$, $|\hat{P}(z, \bar{z})| > 0$ in $|z| \leq \frac{\epsilon}{2}$, and $P_\epsilon(z, \bar{z})$ is in the closed convex hull of $\{e^{i(1-n)\theta} P(z);\ \frac{\epsilon}{2} \leq |z| \leq \epsilon\}$. We define a function f on the complex plane by

$$f(z) = \begin{cases} z - M^{-1}\hat{P}(z, \bar{z}), & |z| \leq 1, \\ z - M^{-1} z^{1-n} P(z), & |z| > 1. \end{cases}$$

At this time, f is continuous, and f maps K into K. In fact, whenever $|z| \leq 1$,

$$\begin{aligned} |f(z)| &\leq |z| + M^{-1}|\hat{P}(z, \bar{z})| \leq 1 + M^{-1}(1 + |a_0| + \cdots + |a_{n-1}|) \\ &\leq 1 + 1 = 2 < M; \end{aligned}$$

whenever $1 \leq |z| \leq M$,

$$\begin{aligned}|f(z)| &= |z - M^{-1}z - M^{-1}z^{1-n}(a_0 + \cdots + a_{n-1}z^{n-1})| \\ &\leq (M-1) + M^{-1}(|a_0| + \cdots + |a_{n-1}|) \\ &\leq M - 1 + M^{-1}(M-2) \leq M.\end{aligned}$$

Since K is a bounded closed convex subset, we know from the Brouwer fixed point theorem that f has at least one fixed point in K, which implies that $P(z_0) = 0$. This is the end.

Theorem 2.14 (Borsuk Theorem). *We say that Ω is* **symmetric** *with respect to the origin 0, if $\Omega = -\Omega$. Let $\Omega \subset \mathbb{R}^n$ be a bounded open subset and symmetric with respect to the origin 0, and $0 \in \Omega$. Suppose that $f \in C(\Omega, \mathbb{R}^n)$ and*

$$f(x) = -f(-x) \neq 0$$

on $x \in \partial\Omega$. Then $\deg(f, \Omega, 0)$ is equal to odd number.

Proof. Because the Brouwer degree is only related to boundary values of a function, let us assume that f is odd in whole $\bar{\Omega}$, that is,

$$f(x) = -f(-x), \quad \forall x \in \bar{\Omega}.$$

By utilizing small perturbation invariance, it can be assumed that $f \in C^1(\bar{\Omega}, \mathbb{R}^n)$ and $J_f(0) \neq 0$. Indeed, assume $g_1 \in C^1(\bar{\Omega}, \mathbb{R}^n)$ such that $||f - g_1||_C$ is sufficiently small. Let

$$g_2(x) = \frac{1}{2}(g_1(x) - g_1(-x)),$$

take δ small enough such that δ is no eigenvalue of $g_2'(0)$. Then $\tilde{f} := g_2 - \delta\, id \in C^1(\bar{\Omega}, \mathbb{R}^n)$, $\tilde{f}$ is odd, $J_{\tilde{f}}(0) \neq 0$, and $||\tilde{f} - f||_C$ is sufficiently small.

To prove the theorem, it suffices to prove that there exists an odd C^1 mapping g such that $||f - g||_C$ is sufficiently small and 0 is the regular value of g.

In fact, if such g exists, then

$$\deg(f, \Omega, 0) = \deg(g, \Omega, 0) = \mathrm{sign} J_g(0) + \sum_{0 \neq x \in g^{-1}(0)} \mathrm{sign} J_g(x).$$

Since $g(x) = 0 \Longleftrightarrow g(-x) = 0$ and $J_g(x) = J_g(-x)$, we obtain that the right-hand side of the above equation is odd.

In the sequel, we will show the existence of such g by induction. Put

$$\Omega_k = \{x = (x_1, \cdots, x_n) \in \Omega;\ \text{there exists some } i \leq k, \\ \text{such that } x_i \neq 0\},$$

and take an odd function $\phi \in C^1(\mathbb{R}, \mathbb{R})$ such that $\phi'(0) = 0$ and $\phi(t) = 0 \Longleftrightarrow t = 0$.

Consider the C^1 mapping $\bar{f}(x) = \frac{f(x)}{\phi(x_1)}$ on the bounded open subset $\Omega_1 = \{x \in \Omega;\ x_1 \neq 0\}$. By Sard Theorem, $\bar{f}$ has a regular value $y^1 \in \mathbb{R}^n$ and $||y^1||$ is sufficiently small as necessary in the sequel.

Let $g_1(x) = f(x) - \phi(x_1)y^1$, $y^1 = (y_1^1, \ldots, y_n^1)^T$. Then $g_1(x) = \phi(x_1)(\bar{f}(x) - y^1)$. Thus,

$$g_1'(x) = \phi(x_1)\bar{f}'(x) + \begin{pmatrix} \phi'(x_1)(\bar{f}_1(x) - y_1^1) & 0 & \cdots & 0 \\ \vdots & \vdots & & \vdots \\ \phi'(x_1)(\bar{f}_n(x) - y_n^1) & 0 & \cdots & 0 \end{pmatrix}.$$

Hence, whenever $x \in \Omega_1$ and $g_1(x) = 0$ (i.e. $\bar{f}(x) = y^1$), we have $g_1'(x) = \phi(x_1)\bar{f}'(x)$. Since y^1 is the regular value of $\bar{f}(x)$, it follows that 0 is a regular value of g_1 defined on Ω_1.

Now, suppose that when $k < n$, there exists some odd C^1 mapping g_k such that $||g_k - f||_C$ is sufficiently small and 0 is a regular value of g_k defined on Ω_k.

Denote

$$g_{k+1}(x) = g_k(x) - \phi(x_{k+1})y^{k+1},$$

where $y^{k+1} \in \mathbb{R}^n$ is a regular value of $\bar{g}_k(x) = \frac{g_k(x)}{\phi(x_{k+1})}$ defined on $\Omega'_{k+1} = \{x \in \Omega;\ x_{k+1} \neq 0\}$, and $||y^{k+1}||$ is sufficiently small.

By using the same method as in the first step, we obtain that whenever $x \in \Omega'_{k+1}$ and $g_{k+1}(x) = 0$ (i.e., $\bar{g}_k(x) = y^{k+1}$), $g'_{k+1}(x) = \phi(x_{k+1})\bar{g}'_k(x)$. Thus, 0 is a regular value of g_{k+1} defined on Ω'_{k+1}.

Additionally, whenever $x \in \Omega_{k+1}$, $x_{k+1} = 0$, and $g_{k+1}(x) = 0$, we have $x \in \Omega_k$ and $g_k(x) = 0, g'_k(x) = g'_{k+1}(x)$ from the properties of ϕ. By induction assumption, $J_{g_{k+1}}(x) = J_{g_k}(x) \neq 0$. Thus, 0 is a regular value of $g_{k+1}(x)$ defined on Ω_k. Since $\Omega_{k+1} = \Omega_k \cup \Omega'_{k+1}$, it follows that 0 is a regular value of g_{k+1} defined on Ω_{k+1}. Evidently, g_{k+1} is odd and close to f on $\bar{\Omega}$.

By induction, we finally obtain that $g = g_n$ is an odd C^1 mapping, $||g - f||_C$ is sufficiently small, and 0 is a regular value of g defined on $\Omega_n = \Omega\backslash\{0\}$. Furthermore, from the properties of ϕ, it follows that $g'(0) = g'_k(0) = f'(0)$, $k = 1, 2, \ldots, n$, and $J_f(0) \neq 0$. Therefore, 0 is a regular value of g defined on the whole Ω. This is the end.

Corollary 2.1. *Let $\Omega \subset \mathbb{R}^n$ be a bounded open subset with $0 \in \Omega$ and symmetric with respect to 0. Suppose that $f \in C(\bar{\Omega}, \mathbb{R}^n)$ and $f(x) \neq tf(-x)$ for any $x \in \partial\Omega$, $t \in [0,1]$. Then* $\deg(f, \Omega, 0)$ *is odd.*

Proof. Let

$$h(t, x) = f(x) - tf(-x), \quad x \in \bar{\Omega},\ t \in [0,1].$$

Then, $h(t,x)$ is continuous, $h(1,x)$ is odd, and $0 \notin h(t, \partial\Omega)$ for any $t \in [0,1]$. By the homotopy invariance and Borsuk Theorem, we arrive at

$$\deg(f, \Omega, 0) = \deg(h_0, \Omega, 0) = \deg(h_1, \Omega, 0) = \text{odd}.$$

This is the end.

Corollary 2.2 (Borsuk–Ulam). *Let $\Omega \subset \mathbb{R}^n$ be a bounded open subset with $0 \in \Omega$ and symmetric with respect to 0. Suppose that $f : \partial\Omega \to \mathbb{R}^m$ is continuous, and $m < n$. Then, there exists $x \in \partial\Omega$ such that $f(x) = f(-x)$. Additionally, if f is still odd, then f has at least one zero point on $\partial\Omega$, i.e. $0 \in f(\partial\Omega)$.*

Proof. We argue it by contradiction. Suppose that $f(x) \neq f(-x)$ on $x \in \partial\Omega$. Let

$$g(x) = f(x) - f(-x).$$

Then g is odd, and $0 \notin g(\partial\Omega)$. By the continuous extension theorem, let g be any continuous extension to $\bar{\Omega}$ of these boundary values such that $g(\bar{\Omega}) \subset \mathbb{R}^m$. On the one hand, $\deg(g, \Omega, 0) = \text{odd}$, from Borsuk Theorem. On the other hand, $\deg(g, \Omega, 0) = 0$, from the lack of direction of Brouwer degree. This is a contradiction. This is the end.

Remark. The metereological interpretation for the above corollary is that at two opposite ends of the earth we have the same weather, i.e. temperature and pressure ($n = 3$ and $m = 2$).

Theorem 2.15. *Let $\Omega \subset \mathbb{R}^n$ be an open subset. Suppose that $f : \Omega \to \mathbb{R}^n$ is continuous and locally one-to-one. Then f is an open mapping.*

Proof. For any fixed point $x_0 \in \Omega$, we will show that $f(x_0)$ is an interior point of $f(\Omega)$. Without loss of generality, it can be assumed that $x_0 = 0, f(x_0) = 0$, if necessary, passing to $\Omega - x_0$ and $f(x + x_0) - f(x_0)$ for $x \in \Omega - x_0$. Take an ϵ-neighborhood $B(0, \epsilon)$ of 0 such that f is one-to-one on $\overline{B(0, \epsilon)}$. Define a homotopy

$$H(x,t) = f\left(\frac{x}{1+t}\right) - f\left(\frac{-tx}{1+t}\right).$$

Then $H : \overline{B(0, \epsilon)} \times [0, 1] \to \mathbb{R}^n$ is continuous, and $H(x, t) \neq 0$ whenever $x \in \partial B(0, \epsilon)$ and $t \in [0, 1]$. This is because

$$\frac{1}{1+t}x \neq \frac{-t}{1+t}x$$

for $x \neq 0$, and f is one-to-one on $\overline{B(0, \epsilon)}$. By the homotopy invariance and Borsuk Theorem, we arrive at

$$\begin{aligned}\deg(f, B(0, \epsilon), 0) &= \deg(H(\cdot, 0), B(0, \epsilon), 0)\\ &= \deg(H(\cdot, 1), B(0, \epsilon), 0) = \text{odd}.\end{aligned}$$

By the property of connected components, if $p \in \mathbb{R}^n$ and $||p||$ is sufficiently small, we obtain

$$\deg(f, B(0, \epsilon), p) = \deg(f, B(0, \epsilon), 0) \neq 0.$$

Hence, if $||p||$ is sufficiently small, $f(x) = p$ admits at least one solution in $B(0, \epsilon)$. Therefore, 0 is an interior point of $f(B(0, \epsilon))$. This is the end.

2.4 Leray–Schauder Degree

One of the important motivation for degree theory is to explain the existence of solution for an equation $f(x) = 0$. Brouwer degree theory is established for continuous functions defined on finite dimensional spaces. Brouwer fixed point theorem has many important applications. Its establishment only need to use the three fundamental properties of the degree theory: normality, additivity, homotopy invariance. In the study of differential equations or integral equations, we need to consider a problem on infinite dimensional Banach spaces. It is well known that Brouwer fixed point theorem cannot be directly extended to infinite dimensional spaces. For example, let $X = c_0$ be the Banach space of real sequences $x = \{x_n\}$, $x_n \to 0$, $||x|| = \max_n |x_n|$. Define $F : X \to X$ by

$$(Fx)_1 = (1 + ||x||)/2, \quad (Fx)_{n+1} = x_n, \quad \forall n \geq 1.$$

Since $||Fx - Fy|| = ||x - y||$, we know that F is continuous and $F : \bar{B}_1(0) \to \bar{B}_1(0)$. But, F has no fixed points. This is because $x = Fx$ implies $x_n = (1 + ||x||)/2$, $\forall n \geq 1$, which implies $x = (x_n) \notin c_0$. This is a contradiction.

The above example shows that one cannot define a topological degree for all continuous mappings defined on general Banach spaces. In 1934, Leray and Schauder considered a class of mappings, established the topological degree (called Leray–Schauder degree) on infinite dimensional spaces, and obtained the Leray–Schauder fixed point theorem.

2.4.1 Compact Continuous Mappings and Its Properties

Let X, Y be normed linear spaces, $D \subset X$.

Theorem 2.16. *Suppose that Y is a Banach space, $F_n : D \to Y$, $n = 1, 2, \ldots$ is a sequence of completely continuous mappings. Assume that for any bounded subset S in D and any $\epsilon > 0$, there is a natural number N such that whenever $n, m \geq N$,*

$$\sup_{x \in S} ||F_n(x) - F_m(x)|| < \epsilon. \tag{2.19}$$

Then there exists a completely continuous mapping $F : D \to Y$ such that $||F_n(x) - F(x)||$ tends to zero uniformly on bounded subsets, as $n \to \infty$.

Proof. Firstly, from (2.19), it is known that for every $x \in D$, $\{F_n(x)\}$ is a Cauchy sequence in Y. Thus, the sequence $\{F_n(x)\}$ converges a point in Y. Denote $F : D \to Y$ by

$$F(x) := \lim_{n\to\infty} F_n(x).$$

By (2.19), $||F_n(x) - F(x)|| \to 0$ uniformly on $x \in S$, as $n \to \infty$. Hence, F is a continuous mapping.

Next, we will show that F is a completely continuous mapping. Suppose that $S \subset D$ is a bounded subset. Since F_n converges to F uniformly on S as $n \to \infty$, it follows that for any $\epsilon > 0$, there is a natural number n such that

$$\sup_{x\in S} ||F_n(x) - F(x)|| < \epsilon/2.$$

Because $F_n(S)$ is relatively compact in Y, it follows that $F_n(S)$ has a finite $\epsilon/2$-net. Thus, $F(S)$ admits a finite ϵ-net. Hence, $F(S)$ is a relatively compact subset in Y. This is the end.

Theorem 2.17. *Suppose that $M \subset X$ is a bounded closed subset and $F : M \to Y$ is continuous. Then the necessary condition for F to be completely continuous is that, for any $\epsilon > 0$, there exist bounded continuous mappings $F_n : M \to Y^n$ with finite dimensional range, such that*

$$\sup_{x\in M} ||F(x) - F_n(x)|| < \epsilon, \tag{2.20}$$

where $Y^n \subset Y$ is a finite dimensional subspace.

Proof. Since $F(M)$ is relatively compact in Y, there exists a finite ϵ-net, that is, for any $\epsilon > 0$, there exist $y_1, y_2, \ldots, y_m \in Y$ such that $\{y_1, y_2, \ldots, y_m\}$ is a finite ϵ-net of $F(M)$, i.e. for any $y \in F(M)$, there is some y_k, $1 \le k \le m$, such that $||y - y_k|| < \epsilon$.

Let Y^n be the n-dimensional subspace spanned by $y_1, \ldots, y_m$. Set

$$B_i = B(y_i, \epsilon) = \{y \in Y;\ ||y - y_i|| < \epsilon\}, \quad i = 1, 2, \ldots, m.$$

Define the functions $d_i : Y \to \mathbb{R}$ by

$$d_i(y) = \mathrm{dist}(y, Y \backslash B_i), \quad i = 1, 2, \ldots, m.$$

Denote $F_n : M \to Y^n$ by

$$F_n(x) = \frac{1}{\sum_{i=1}^m d_i(F(x))} \sum_{i=1}^m d_i(F(x))y_i, \quad x \in M.$$

Since d_i is nonnegative, continuous, and positive in the interior of B_i, and $F(M) \subset \bigcup_{i=1}^m B_i$, it follows that $\sum_{i=1}^m d_i(F(x)) > 0$ for any $x \in M$. Thus, the mapping F_n is well defined and continuous. Because $d_i(F(x)) = 0$ whenever $||F(x) - y_i|| \geq \epsilon$, we arrive at

$$\begin{aligned} ||F(x) - F_n(x)|| &= \frac{1}{\sum_{i=1}^m d_i(F(x))} \left\| \sum_{i=1}^m d_i(F(x))(F(x) - y_i) \right\| \\ &\leq \frac{1}{\sum_{i=1}^m d_i(F(x))} \sum_{i=1}^m d_i(F(x))||F(x) - y_i|| < \epsilon. \end{aligned}$$

This completes the proof.

Remark. Due to the bounded continuity and finite dimensional range of F_n, F_n is completely continuous. Using Theorem 2.16, when Y is complete, the conditions for Theorem 2.17 are still sufficient.

Theorem 2.18. *Let X be a normed linear space, Y a Banach space, $U \subset X$ open subset, and $x_0 \in U$. Suppose that $F \in C^m(U, Y)$, $m \geq 1$ and F is completely continuous. Then $F^{(k)}(x_0), k = 1, 2, \ldots, m$, map the bounded subsets $\left\{(\underbrace{h, \ldots, h}_{k}); h \in X, ||h|| \leq r\right\}$ to relatively compact subsets in Y. Especially, $F'(x_0)$ is a completely continuous linear operator.*

Proof. Since $F^{(k)}(x_0) \in L_s^{(k)}(X, Y)$, $F^{(k)}(x_0)$ is continuous. Using homogeneity, it suffices to prove that $F^{(k)}(x_0)$ maps $B = \{h^{(k)};\ h \in X, ||h|| \leq 1\}$ to a relatively compact subset in Y.

When $k = 0$, the conclusion is correct.

Using induction, we assume that the conclusion holds when $n = 1, \ldots, k-1$ ($k \leq m$). We consider $F^{(k)}(x_0)$.

By using the argument of contradiction, suppose that $F^{(k)}(x_0)$ would not be compact on B. Then, there would be $\epsilon > 0$ and a sequence of $\{h_i^{(k)}\}$ in B, such that, when $i \neq j$,

$$\frac{1}{k!}||F^{(k)}(x_0)h_i^{(k)} - F^{(k)}(x_0)h_j^{(k)}|| \geq \epsilon.$$

Using Taylor's formula, we have

$$F(x_0 + h) = \sum_{n=0}^{k-1} \frac{1}{n!} F^{(n)}(x_0)h^{(n)} + \frac{1}{k!}F^{(k)}(x_0)h^{(k)} + w(x_0, h),$$

where

$$\lim_{||h|| \to 0} \frac{||w(x_0, h)||}{||h||^k} = 0. \tag{2.21}$$

From (2.21), we obtain that, for sufficient small $\rho > 0$,

$$||w(x_0, \rho h)|| < \rho^k \frac{\epsilon}{k+3}.$$

By using inductive assumptions and selecting subsequences, it is possible to assume that $\{h_i\}$ satisfies that, for $0 \leq n < k$ and sufficient large i, j,

$$\frac{1}{n!}||F^{(n)}(x_0)h_i^{(n)} - F^{(n)}(x_0)h_j^{(n)}|| < \rho^{k-n} \frac{\epsilon}{k+3}.$$

Thus, when i, j are sufficient large, we arrive at

$$\begin{aligned}
&||F(x_0 + \rho h_i) - F(x_0 + \rho h_j)|| \\
&\quad \geq \frac{1}{k!}||F^{(k)}(x_0)h_i^{(k)} - F^{(k)}(x_0)h_j^{(k)}||\rho^k \\
&\qquad - \sum_{n=0}^{k-1} \frac{1}{n!}||F^{(n)}(x_0)h_i^{(n)} - F^{(n)}(x_0)h_j^{(n)}||\rho^n \\
&\qquad - ||w(x_0, \rho h_i)|| - ||w(x_0, \rho h_j)|| \\
&\quad \geq \epsilon \rho^k - \sum_{n=0}^{k-1} \frac{\epsilon}{k+3}\rho^k - \frac{2\epsilon}{k+3}\rho^k \geq \frac{\epsilon}{k+3}\rho^k > 0.
\end{aligned}$$

This shows that $\{F(x_0 + \rho h_i)\}$ is not relatively compact, which is a contradiction. This completes the proof.

Example 2.4. Let $\Omega \subset \mathbb{R}^n$ be a bounded closed subset and $k(x,y,u)$ a continuous function on $\Omega \times \Omega \times \mathbb{R}$. Then the integral operator $K : C(\Omega) \to C(\Omega)$,

$$(K\phi)(x) = \int_\Omega k(x,y,\phi(y))dy$$

is a completely continuous operator.

Proof. Assume that B is a bounded subset in $C(\Omega)$, we will show that $K(B)$ is a relatively compact subset in $C(\Omega)$. By using the Arzela–Ascoli's Theorem, it suffices to show that $K(B)$ is uniformly bounded and equi-continuous.

Since B is a bounded subset, there exists N such that $||\phi||_C \leq N$, $\forall \phi \in B$. Due to the continuity of $k(x,y,u)$, there is an $M > 0$ such that $|k(x,y,u)| \leq M$, $\forall (x,y,u) \in \Omega \times \Omega \times [-N,N]$. Thus, we arrive at

$$|K\phi(x)| = \left| \int_\Omega k(x,y,\phi(y))dy \right| \leq M \text{mes}\, \Omega, \quad \forall \phi \in B.$$

Hence, $K(B)$ is uniformly bounded.

Since $k(x,y,u)$ is uniformly continuous on $\Omega \times \Omega \times [-N,N]$, it implies that, $\forall \epsilon > 0$, there exists $\delta > 0$, such that, whenever $|x_1 - x_2| < \delta$ $(x_1, x_2 \in \Omega)$,

$$|k(x_1,y,u) - k(x_2,y,u)| < \frac{\epsilon}{1 + \text{mes}\, \Omega}, \quad \forall y \in \Omega, |u| \leq N.$$

Then, whenever $|x_1 - x_2| < \delta$, $\forall \phi \in B$,

$$|K\phi(x_1) - K\phi(x_2)| = \left| \int_\Omega [k(x_1,y,\phi(y)) - k(x_2,y,\phi(y))]dy \right| < \epsilon.$$

Therefore, $K(B)$ is equi-continuous.

Finally, we will show that K is continuous. Suppose that $\phi_n, \phi_0 \in C(\Omega)$, $||\phi_n - \phi_0||_C \to 0$. Thus, $\{\phi_n\}$ is bounded in $C(\Omega)$. Let $N_1 \geq \sup\{||\phi_0||_C, ||\phi_1||_C, \ldots\}$. Since $k(x,y,u)$ is uniformly continuous on $\Omega \times \Omega \times [-N_1, N_1]$, it follows that, $\forall \epsilon > 0$, there exists $\delta > 0$, such that, whenever $u_1, u_2 \in [-N_1, N_1]$ and $|u_1 - u_2| < \delta$,

$$|k(x,y,u_1) - k(x,y,u_2)| < \frac{\epsilon}{1 + \text{mes}\, \Omega}, \quad \forall (x,y) \in \Omega \times \Omega.$$

Take N such that, whenever $n \geq N$, $||\phi_n - \phi_0||_C < \delta$. Thus,

$$|K\phi_n(x) - K\phi_0(x)| \leq \int_\Omega |k(x, y, \phi_n(y)) - k(x, y, \phi_0(y))|dy < \epsilon.$$

Hence, $||K\phi_n - K\phi_0||_C < \epsilon$. This is the end.

2.4.2 Completely Continuous Field and Compact Homotopy

Definition 2.4.1. Let X be a real normed linear space and $D \subset X$. If $F : D \to X$ is completely continuous, then the mapping $f = id - F$ is said to be **completely continuous field**, or **compact continuous field** (or a completely continuous perturbation of the identity).

Theorem 2.19. *Let $D \subset X$ be a closed bounded subset and $f = id - F : D \to X$ a completely continuous field. Then,*

(1) f is proper, that is, $f^{-1}(K)$ is compact whenever K is compact.
(2) f is closed, that is, $f(S)$ is closed whenever $S \subset D$ is closed.

Proof. (1) Assume $x_n \in f^{-1}(K)$. Then, let $y_n = f(x_n) = x_n - F(x_n) \in K$. Since K is compact, we can assume $y_n \to y \in K$, without loss of generality. Since F is compact and D is bounded, we can assume $F(x_n) \to z$ (we take a subsequence if it is necessary). Thus, $x_n = y_n + F(x_n) \to y + z =: x$. Since F is continuous and $y_n = x_n - F(x_n)$, it follows that $y = x - F(x) = f(x)$, that is, $x \in f^{-1}(y) \subset f^{-1}(K)$. Hence, $f^{-1}(K)$ is compact.

(2) Assume that $z_n \in f(S)$, $z_n \to z_0$. We want to show $z_0 \in f(S)$. By definition, there exist $x_n \in S$ such that $z_n = f(x_n) = x_n - F(x_n)$. Since F is completely continuous, there is a subsequence $F(x_{n_k}) \to y_0$. Thus, $x_{n_k} = z_{n_k} + F(x_{n_k}) \to z_0 + y_0 =: x_0 \in S$. From the continuity of F, it implies that $z_0 = x_0 - F(x_0) = f(x_0)$. Hence, $z_0 \in f(S)$. This completes the proof.

Corollary 2.3. *Let $M \subset X$ be a closed bounded subset and $f = id - F : M \to X$ a completely continuous field. If $p \in X \backslash f(M)$, then* $\text{dist}(p, f(M)) = \epsilon > 0$.

Definition 2.4.2. Let $M \subset X$ be a closed bounded subset, $f_0 = id - F_0$ and $f_1 = id - F_1$ be two compact continuous fields on M. We say that the two compact continuous fields $f_0 = id - F_0$ and $f_1 = id - F_1$ are **compact homotopy** on M, if there is a completely continuous mapping $H : M \times [0,1] \to X$ such that, whenever $x \in M$ and $t \in [0,1]$,

$$h_t(x) = x - H(x,t) \neq 0, \quad H(x,0) = F_0(x), \quad H(x,1) = F_1(x).$$

Remark. The compactness of H in compact homotopy refers to the compactness with respect to two variables, that is, $\overline{H(M \times [0,1])}$ is a compact subset. For an infinite dimensional space, even if H is continuous with two variables and $H(x,t)$ is completely continuous with respect to x for any fixed $t \in [0,1]$, it cannot be inferred that H is compact with respect to two variables. But we have the following conclusion:

Suppose that $H : \bar{\Omega} \times [0,1] \to X$ is continuous; for each fixed $t \in [0,1]$, $H(\cdot,t) : \bar{\Omega} \to X$ is compact continuous; and the continuity of $H(x,t)$ at any point $t_0 \in [0,1]$ is uniform about $x \in \bar{\Omega}$. Then $H : \bar{\Omega} \times [0,1] \to X$ is completely continuous.

Please see [13] for its proof.

2.4.3 The Definition of Leray–Schauder Degree

Let X be a real normed linear space, $\Omega \subset X$ be an open bounded subset, $F : \bar{\Omega} \to X$ be completely continuous, $f = id - F$ be a completely continuous field, and $p \notin f(\partial\Omega)$.

By Corollary 2.3, there exists $\epsilon > 0$, such that

$$\operatorname{dist}(p, f(\partial\Omega)) = \epsilon > 0.$$

By Theorem 2.17, there exist a finite dimensional subspace X_n of X with $p \in X_n$, and a bounded continuous mapping $F_n : \bar{\Omega} \to X_n$ such that

$$\sup_{x \in \bar{\Omega}} ||F(x) - F_n(x)|| < \epsilon.$$

Let $\Omega_n = X_n \cap \Omega$ and consider the following finite dimensional continuous mapping

$$f_n(x) = x - F_n(x), \quad x \in \bar{\Omega}_n,$$

Then, whenever $x \in \partial\Omega_n (\subset \partial\Omega)$, we have

$$||f_n(x) - p|| \geq ||f(x) - p|| - ||F_n(x) - F(x)|| > 0,$$

that is, $p \notin f_n(\partial\Omega_n)$. Thus, the Brouwer degree $\deg(f_n, \Omega_n, p)$ is well defined. We use this as the definition of the Leray–Schauder degree of f on Ω with respect to the point p.

Definition 2.4.3. Let X_n be a finite dimensional subspace of X, $p \in X_n$, and $F_n : \bar{\Omega} \to X_n$ be continuous, such that

$$\sup_{x \in \bar{\Omega}} ||F(x) - F_n(x)|| < \epsilon = \text{dist}(p, f(\partial\Omega)).$$

Then the **Leray–Schauder degree** of the completely continuous field f on Ω with respect to p is defined by

$$\deg(f, \Omega, p) = \deg(f_n, \Omega_n, p),$$

where $f = id - F$, $f_n = id - F_n$.

The following lemma illustrates the validity of this definition.

Lemma 2.11. *Let X_n be any finite dimensional subspace of X with $p \in X_n$, and $F_n : \bar{\Omega} \to X_n$ be continuous, such that*

$$\sup_{x \in \bar{\Omega}} ||F(x) - F_n(x)|| < \epsilon = \text{dist}(p, f(\partial\Omega)).$$

Then $\deg(f_n, \Omega_n, p)$ is independent of the choice of X_n, F_n.

Proof. Firstly, we prove that when X_n is fixed, $\deg(f_n, \Omega_n, p)$ is independent of the choice of F_n.

Suppose that $F_n, G_n : \bar{\Omega} \to X_n$ are continuous, such that

$$\sup_{x \in \bar{\Omega}} ||F(x) - F_n(x)|| < \epsilon, \quad \sup_{x \in \bar{\Omega}} ||F(x) - G_n(x)|| < \epsilon.$$

Let $f_n = id - F_n$, $g_n = id - G_n$. Define a homotopy

$$h_t(x) := x - tF_n(x) - (1-t)G_n(x), \quad x \in \bar{\Omega}, \ t \in [0,1].$$

Then, whenever $x \in \partial\Omega$ and $t \in [0,1]$, we have

$$||h_t(x) - p|| \geq ||f(x) - p|| - t||F(x) - F_n(x)|| - (1-t)||G_n(x) - F(x)|| > 0.$$

According to the homotopy invariance of Brouwer degree, we arrive at

$$\deg(f_n, \Omega_n, p) = \deg(g_n, \Omega_n, p).$$

Next, we prove that $\deg(f_n, \Omega_n, p)$ is independent of the choice of X_n.

Suppose that X_m is any finite dimensional subspace of X and $X_n \subset X_m$. From the above proof, $\deg(f_m, \Omega_m, p)$ is independent of the choice of F_m. Specifically, we set that F_m is F_n. Then,

$$\deg(f_m, \Omega_m, p) = \deg(f_n, \Omega_m, p).$$

According to the reduction theorem of Brouwer degree, we obtain

$$\deg(f_n, \Omega_m, p) = \deg(f_n\big|_{X_n \cap \Omega_m}, \Omega_m \cap X_n, p) = \deg(f_n, \Omega_n, p).$$

If X_n and X_k are any finite dimensional subspaces, we denote X_m as the subspace spanned by X_n and X_k. From the above proof, we arrive at

$$\deg(f_n, \Omega_n, p) = \deg(f_m, \Omega_m, p) = \deg(f_k, \Omega_k, p).$$

This completes the proof.

2.4.4 Properties of the Leray–Schauder Degree

Let $\Omega \subset X$ be an open bounded subset, and $f = id - F$ be a completely continuous field on $\bar{\Omega}$.

Theorem 2.20. *The Leray–Schauder degree enjoys the following fundamental properties:*

(1) (**Normality**)

$$\deg(id, \Omega, p) = \begin{cases} 1, & p \in \Omega, \\ 0, & p \notin \bar{\Omega}. \end{cases}$$

(2) (**Additivity**) *If Ω_1, Ω_2 are two disjoint open subsets of Ω and $p \notin f(\bar{\Omega} \backslash (\Omega_1 \cup \Omega_2))$, we have*

$$\deg(f, \Omega, p) = \deg(f, \Omega_1, p) + \deg(f, \Omega_2, p).$$

(3) (**Homotopy invariance**) *Let* $H : \bar{\Omega} \times [0,1] \to X$ *be compact continuous, and* $h_t(x) = x - H(x,t)$*. Suppose that* $p : [0,1] \to X$ *is continuous, and* $p(t) \notin h_t(\partial\Omega)$ *as* $t \in [0,1]$*. Then,*

$$\deg(h_t, \Omega, p(t)) \text{ is independent of } t.$$

Proof. (1) It is obvious.

(2) It is easy to see that $\Omega_0 = \bar{\Omega} \setminus (\Omega_1 \cup \Omega_2)$ is closed. By $p \notin f(\Omega_0)$ and Corollary 2.3, there exists $\epsilon > 0$ such that

$$\operatorname{dist}(p, f(\Omega_0)) \geq \epsilon.$$

Take X_n and $F_n : \bar{\Omega} \to X_n$ such that $p \in X_n$ and

$$\sup_{x \in \bar{\Omega}} ||F(x) - F_n(x)|| < \epsilon.$$

Let $f_n = id - F_n$. From Definition 2.4.3, we obtain

$$\begin{aligned}
\deg(f, \Omega, p) &= \deg(f_n, \Omega \cap X_n, p),\\
\deg(f, \Omega_1, p) &= \deg(f_n, \Omega_1 \cap X_n, p),\\
\deg(f, \Omega_2, p) &= \deg(f_n, \Omega_2 \cap X_n, p).
\end{aligned}$$

Since $p \notin f_n(\Omega_0 \cap X_n)$, it follows from the additivity of the Brouwer degree that

$$\deg(f_n, \Omega \cap X_n, p) = \deg(f_n, \Omega_1 \cap X_n, p) + \deg(f_n, \Omega_2 \cap X_n, p).$$

Consequently,

$$\deg(f, \Omega, p) = \deg(f, \Omega_1, p) + \deg(f, \Omega_2, p).$$

(3) From the definition, we know that if $g(x) = f(x) - p$, $\deg(f, \Omega, p) = \deg(g, \Omega, 0)$. Thus, we can assume $p(t) = p$ without loss of generality. Since H is compact continuous with respect to two variables, it follows from the assumption $p \notin h_t(\partial\Omega)$ that there exists $\epsilon > 0$, such that

$$\inf_{x \in \partial\Omega, t \in [0,1]} ||p - h_t(x)|| \geq \epsilon.$$

Since $H : \bar{\Omega} \times [0,1] \to X$ is compact continuous, it follows from Theorem 2.17 that there exist a finite dimensional subspace X_n and a continuous mapping $H_n : \bar{\Omega} \times [0,1] \to X_n$ such that $p \in X_n$ and

$$\sup_{x\in\bar{\Omega},t\in[0,1]} ||H(x,t) - H_n(x,t)|| < \epsilon.$$

Set

$$h_{nt}(x) = x - H_n(x,t).$$

Then, if $x \in \partial\Omega$ and $t \in [0,1]$, we have

$$||h_{nt}(x) - p|| \geq ||h_t(x) - p|| - ||h_t(x) - h_{nt}(x)|| > 0,$$

that is, $p \notin h_{nt}(\partial\Omega)$. From the homotopy invariance of the Brouwer degree, we obtain that $\deg(h_{nt}, \Omega \cap X_n, p)$ is independent of t. Hence, it follows from the definition of the Leray–Schauder degree that

$$\deg(h_t, \Omega, p) = \deg(h_{nt}, \Omega \cap X_n, p).$$

This completes the proof.

Theorem 2.21. *For completely continuous field $f = id - F$, the Leray–Schauder degree* $\deg(f, \Omega, p)$ *possesses the following properties*:

(1) (**Excision**) *If $K \subset \bar{\Omega}$ is a closed subset and $p \notin f(K)$, then*

$$\deg(f, \Omega, p) = \deg(f, \Omega\backslash K, p).$$

(2) (**Kronecker existence**) *If $p \notin f(\bar{\Omega})$, then* $\deg(f, \Omega, p) = 0$. *If* $\deg(f, \Omega, p) \neq 0$, *then the equation $f(x) = p$ admits at least one solution in Ω.*

(3) (**Connected components**) *When p moves within a connected component of $X\backslash f(\partial\Omega)$,* $\deg(f, \Omega, p)$ *does not change (is constant). Particularly, if p is in unbounded component,* $\deg(f, \Omega, p) = 0$.

(4) (**Boundary value**) $\deg(f, \Omega, p)$ *only depends on restriction of f on $\partial\Omega$, i.e., if $g : \bar{\Omega} \to X$ is any completely continuous field and $g\big|_{\partial\Omega} = f\big|_{\partial\Omega}$, then*

$$\deg(f, \Omega, p) = \deg(g, \Omega, p).$$

(5) (**Poincaré–Bohl**) *Let $f = id - F, g = id - G$ be completely continuous fields. Suppose that p is not on the line segment connected by $f(x)$ and $g(x)$ for any $x \in \partial\Omega$. Then*

$$\deg(f, \Omega, p) = \deg(g, \Omega, p).$$

(6) (**Acute-angle principle**) *Let* X *be a Hilbert space, and* $0 \in \Omega \subset X$. *Suppose that whenever* $x \in \partial\Omega$, *the inner product* $\langle x, f(x)\rangle > 0$. *Then* $\deg(f, \Omega, 0) = 1$.
(7) (**Lack of direction**) *Suppose that* $0 \notin f(\partial\Omega)$ *and there exists a fixed* $x_0 \in X\backslash\{0\}$ *such that, whenever* $x \in \partial\Omega$ *and* $\lambda > 0$, $f(x) \neq \lambda x_0$. *Then* $\deg(f, \Omega, 0) = 0$.

The proof is easy. Please readers to provide it themselves.

Theorem 2.22 (Product Formula). *Let* Ω, D *be open bounded subsets of* X, $F : \bar{\Omega} \to X$ *be a completely continuous,* $f = id - F$, $f(\bar{\Omega}) \subset D$. *Suppose that* $G : \bar{D} \to X$ *is completely continuous,* $g = id - G$, D_α *are the bounded connected components of* $D\backslash f(\partial\Omega)$, *and* $p \notin (g \circ f)(\partial\Omega) \cup g(\partial D)$. *Then*

$$\deg(g \circ f, \Omega, p) = \sum_{\alpha} \deg(g, D_\alpha, p) \deg(f, \Omega, D_\alpha),$$

where $\deg(f, \Omega, D_\alpha) = \deg(f, \Omega, q)$, $q \in D_\alpha$, *is independent of the choice of* q *from the properties of connected components.*

The proof of this theorem is relatively long, and interested readers can refer to other books [3, 10, 13, 28].

2.4.5 Index of Isolated Zeros

Let X be a real Banach space, $\Omega \subset X$ be an open bounded subset, and $f = id - F : \bar{\Omega} \to X$ be a completely continuous filed.

Definition 2.4.4. If $x_0 \in \Omega$ is the unique zero point of f, then we call

$$\text{index}(f, x_0) := \deg(f, \Omega, 0)$$

the **index** of this zero point.

According to the excision theorem, the index $\text{index}(f, x_0)$ is only related to f and x_0, and is not related to the selection of Ω. It is easy to prove the following theorem.

Theorem 2.23. *Suppose that the completely continuous field* $f = id - F : \bar{\Omega} \to X$ *has only finite zeros* $x_1, \ldots, x_n$ *in* Ω, *and does not* *have any zero on $\partial\Omega$. Then the following index formula holds.*

$$\deg(f, \Omega, 0) = \sum_{i=1}^{n} \text{index}(f, x_i).$$

Theorem 2.24. *Let $f = id - F : \bar{\Omega} \to X$ be a completely continuous field, and x_0 be a zero of f in Ω. Suppose that f is Frechét differentiable at x_0, and 1 is not eigenvalue of $A := F'(x_0)$. Then x_0 is the isolated zero point of f, and*

$$\text{index}(f, x_0) = \text{index}(id - A, 0) = (-1)^{\beta},$$

where β is the sum of the algebraic multiplicities of the eigenvalues μ of the completely continuous operator A satisfying $\mu \in (1, \infty)$.

Proof. By the assumption, $f(x_0) = 0, f'(x_0) = id - A$. According to the definition of F-differentiable, there exists a neighborhood $B(x_0, \epsilon)$ of x_0, such that

$$f(x) = (id - A)(x - x_0) + R(x, x_0),$$

where $R(x, x_0) = o(||x - x_0||)$ is completely continuous operator. Since 1 is not eigenvalue of A and A is completely continuous operator, there exists $m > 0$ such that, whenever $x - x_0 \in X$,

$$||(id - A)(x - x_0)|| \geq m||x - x_0||.$$

Thus, if ϵ is sufficiently small and $||x - x_0|| \leq \epsilon$, we have

$$||R(x, x_0)|| < m||x - x_0||.$$

Hence, x_0 is the isolated zero point of f. Whenever $t \in [0, 1]$ and $0 \neq ||x - x_0|| \leq \epsilon$,

$$\tilde{h}_t(x) := (id - A)(x - x_0) + tR(x, x_0) \neq 0.$$

According to the compact homotopy invariance, we obtain

$$\text{index}(f, x_0) = \text{index}(id - A, 0).$$

In the sequel, we will show $\text{index}(id - A, 0) = (-1)^{\beta}$.

According to the Riesz–Schauder Theory of completely continuous linear operator, the nonzero eigenvalues of A are isolated.

Thus, there are a finite number of eigenvalues greater than 1 for A, and for each nonzero eigenvalue λ of A, the dimension of the invariant subspace

$$N_\lambda = \bigcup_{n=1}^{\infty} \{x \in X;\ (\lambda\, id - A)^n x = 0\}$$

is finite, which is called the **algebraic multiplicity** of λ.

Suppose that all real eigenvalues of A greater than 1 are denoted by $\lambda_1, \ldots, \lambda_k$. Let $X_1 = \bigoplus_{i=1}^{k} N_{\lambda_i}$. From the Riesz–Schauder theory, it follows that there exists an invariant closed subspace X_2, such that

$$X = X_1 \oplus X_2.$$

So, for any $x \in X$, there is a unique decomposition

$$x = x_1 + x_2, \quad x_1 \in X_1,\ x_2 \in X_2.$$

On the unit ball B, put

$$h_t(x) := (2t-1)x_1 + x_2 - tAx = x - H(x,t).$$

Since X_1 is a finite dimensional subspace, it is known that $H(x,t)$ is compact continuous. We claim that whenever $x \in \partial B =: S$ and $t \in [0,1]$, $h_t(x) \neq 0$. In fact, using the method of contradiction, if there are $x = x_1 + x_2 \in S$, $t \in [0,1]$ such that $h_t(x) = 0$, then

$$(2t-1)x_1 - tAx_1 = 0, \quad x_2 - tAx_2 = 0.$$

Since A has no eigenvalues greater than or equal to 1 on X_2, $x_2 = 0$. In addition, since the eigenvalues of A on X_1 are all greater than 1 and $\frac{2t-1}{t} \leq 1$ whenever $t \in (0,1]$, $x_1 = 0$; when $t = 0$, there is also an $x_1 = 0$. This is in contradiction with $x \in S$.

According to the compact homotopy invariance, we obtain

$$\deg(id - A, B, 0) = \deg(h_1, B, 0) = \deg(h_0, B, 0).$$

At this point, $h_0(x) = -x_1 + x_2$. From the definition of Leray–Schauder degree, we obtain

$$\deg(h_0, B, 0) = \deg(-id, X_1 \cap B, 0) = (-1)^\beta.$$

This completes the proof.

2.5 Fixed Point Theorem

Suppose that X is a real Banach space, $\Omega \subset X$ is a bounded open subset, $f = id - F : \bar{\Omega} \to X$ is a completely continuous field, and $\deg(f, \Omega, p)$ is the Leray–Schauder degree.

2.5.1 Leray–Schauder Fixed Point Theorem

Theorem 2.25 (Leray–Schauder). *Let $\Omega \subset X$ be a bounded open subset containing the origin. Suppose that $F : \bar{\Omega} \to X$ is compact continuous and satisfies* **Leray–Schauder boundary condition**: *whenever $x \in \partial\Omega$ and $\lambda > 1$, $F(x) \neq \lambda x$. Then F admits at least one fixed point in $\bar{\Omega}$.*

Proof. Let $f = id - F$. It is sufficient to prove that f must have at least one zero point in $\bar{\Omega}$. Suppose that f have no zeros on $\partial\Omega$, otherwise the conclusion will naturally hold. Denote

$$h_t(x) = x - tF(x), \quad x \in \bar{\Omega}, \ t \in [0, 1],$$

then $h_t(x)$ is a completely continuous field with respect to two variables. According to Leray–Schauder boundary condition, it is known that whenever $x \in \partial\Omega$ and $t \in [0, 1]$, $h_t(x) \neq 0$. By homotopy invariance and normality,

$$\deg(f, \Omega, 0) = \deg(id, \Omega, 0) = 1.$$

From Kronecker existence, f must have at least one zero point in Ω. This is the end.

Remark. The boundary condition can be rewritten as in the form: whenever $x \in \partial\Omega$ and $\mu \in [0, 1)$, $x \neq \mu F(x)$. In order to check such condition, we usually use the prior estimation to do it, that is, we will prove that if the equation $x = \mu F(x)\,(\mu \in [0, 1))$ has a solution x in $\bar{\Omega}$, then x must be inside Ω.

Corollary 2.4 (Leray–Schauder Theorem). *Suppose that $D \subset X$ is a bounded closed convex subset containing at least one interior point, $F : D \to X$ is completely continuous, and $F(\partial D) \subset D$. Then F has at least one fixed point in D.*

Proof. Since D contains at least one interior point, we can assume that 0 is such interior point without loss of generality, otherwise we can make a translation, that is, consider $\Omega - x_0$ and $F(x + x_0) - x_0$ for $x \in \Omega - x_0$ if necessary. Since $F(\partial D) \subset D$ and D is convex, it follows that whenever $x \in \partial D$ and $\lambda > 1$, $F(x) \neq \lambda x$. From Theorem 2.25, it follows that F admits at least one fixed point in D. This is the end.

Corollary 2.5. *Let $\Omega \subset X$ be a bounded open subset, $0 \in \Omega$, and $F : \bar{\Omega} \to X$ be completely continuous. Suppose one of the following condition holds:*

(1) *Rothe condition: if $x \in \partial\Omega$, then $||F(x)|| \leq ||x||$.*
(2) *Krasnosel'skii condition: if X be inner product space and $x \in \partial\Omega$, then $\langle F(x), x\rangle \leq ||x||^2$.*

Then F admits at least one fixed point in $\bar{\Omega}$.

In fact, each one of the conditions (1) and (2) can imply Leray–Schauder boundary condition. From the proof of Theorem 2.25, we know that $\deg(id - F, \Omega, 0) = 1$ under the condition of the above corollary.

In Leray–Schauder Theorem above, there is a condition that Ω possesses an interior point. I would like to remove this condition below. For this, some concepts and conclusions from topology are needed.

Definition 2.5.1. Let A and B be two subsets of the topological space X, $A \subset B$. If there exists a continuous mapping $r : B \to A$ such that $r(x) = x$ whenever $x \in A$, then A is called a **retract** of B and r is called a **retraction**.

We have the following theorem about retract.

Theorem 2.26. *Suppose that $D \subset X$ is a closed convex subset. Then D is a retract of X.*

The proof of Theorem 2.26 can be found in relevant books [14].

Theorem 2.27 (Leray–Schauder Fixed Point Theorem). *Suppose that $D \subset X$ is a bounded closed convex subset, $F : D \to D$ is completely continuous. Then F admits at least one fixed point in D.*

Proof. Let $B_R(0)$ be a ball with radius R centered at 0, so that when R is sufficiently large, $D \subset B_R(0)$. By Theorem 2.26, D is a retract of $B_R(0)$ and its retraction is r. Define $\tilde{F} : B_R(0) \to D$ by

$$\tilde{F} = F \circ r.$$

Then x is a fixed point of $\tilde{F}$ if and only if x is a fixed point of F. It is easy to see that $\tilde{F}$ is compact continuous, and $\tilde{F}(\partial B_R(0)) \subset D \subset B_R(0)$. By Corollary 2.4, $\tilde{F}$ admits at least one fixed point x_0 in $B_R(0)$, that is, $\tilde{F}(x_0) = F(r(x_0)) = x_0 \in D$. From the definition of r, $r(x_0) = x_0$. Hence, $F(x_0) = x_0$. This is the end.

As an application, we will prove a fundamental existence theorem in ordinary differential equations.

Example 2.5 (Peano). Consider the initial value problem

$$(E): \quad \frac{dy}{dx} = f(x, y), \quad y(x_0) = y_0,$$

where $f(x, y)$ is continuous in rectangular region

$$R: \quad |x - x_0| \leq a, \quad ||y - y_0|| \leq b.$$

Then the initial value problem (E) admits at least one solution $y = \phi(x)$ defined on the interval $|x - x_0| \leq h$, where $h = \min\{a, \frac{b}{M}\}$, $M = \max_{(x,y)\in R} ||f(x, y)||$.

Proof. Let $C = C(J, \mathbb{R}^n)$ be the vector space of all continuous vector functions $\varphi(x)$ defined on the interval $J = [x_0 - h, x_0 + h]$. Denote the norm

$$||\varphi|| := \max\{|\varphi(x)|;\ x \in J\}.$$

It is easy to verify that C is a Banach space.

Consider the following subset K in the space C

$$K := \{y(\cdot) \in C;\ ||y(\cdot) - y_0|| \leq Mh\}.$$

Clearly, K is a closed convex subset of C. Define the following mapping on K

$$(Ty)(x) = y_0 + \int_{x_0}^{x} f(s, y(s))ds, \quad y(\cdot) \in K.$$

Then

$$|(Ty)(x) - y_0| = \left| \int_{x_0}^{x} f(s, y(s))ds \right| \leq Mh, \quad x \in J,$$

that is,

$$||(Ty)(\cdot) - y_0|| \leq Mh. \tag{2.22}$$

Thus, $T(K) \subset K$, which indicates that T is an operator of K to itself.

We further prove that T is completely continuous on K. Assume that $y^*(\cdot), y_k(\cdot) \in K(k = 1, 2, \ldots)$ and $y_k(\cdot) \to y^*(\cdot)$, which means that $y_k(x)$ converges to $y^*(x)$ uniformly on J. Since

$$(Ty_k)(x) = y_0 + \int_{x_0}^{x} f(s, y_k(s))ds,$$

we obtain

$$(Ty_k)(x) \to y_0 + \int_{x_0}^{x} f(s, y^*(s))ds = (Ty^*)(x),$$

uniformly on J as $k \to \infty$, which indicates that T is continuous on K. Furthermore, for any $y(\cdot) \in K$ and any $x_1, x_2 \in J$, we have

$$|(Ty)(x_1) - (Ty)(x_2)| = \left| \int_{x_1}^{x_2} f(s, y(s))ds \right| \leq M|x_1 - x_2|.$$

Hence, $T(K)$, as a family of functions defined on J, is equi-continuous on J. Moreover, it can be seen from (2.22) that this family of functions is uniformly bounded on J. By Ascoli–Arzela Theorem, $T(K)$ is relatively compact.

By Leray–Schauder Fixed Point Theorem, there exists a $\phi(\cdot) \in K$, such that $(T\phi)(\cdot) = \phi(\cdot)$, i.e.,

$$\phi(x) = y_0 + \int_{x_0}^{x} f(s, \phi(s))ds, \quad x \in J.$$

Therefore, the initial value problem (E) has at least one solution on interval J. This is the end.

Theorem 2.28 (Krasnosel'skii Fixed Point Theorem). *Let $D \subset X$ be a nonempty bounded closed convex subset. Suppose that T, S are two mappings from D to X satisfying the following conditions:*

(1) *S is continuous and $S(D)$ is included in a compact subset of X.*
(2) *T is a contraction mapping with contraction constant $\alpha < 1$.*
(3) *For any $x, y \in D$, $Tx + Sy \in D$.*

Then there exists $x_0 \in D$ such that $Sx_0 + Tx_0 = x_0$.

Proof. For any $z \in S(D)$, $T+z : D \to D$ is a contraction mapping. Thus, $Tx + z = x$ has a unique fixed point $x = \tau(z) \in D$. So, for any z_1, $z_2 \in S(D)$, we have

$$T(\tau(z_1)) + z_1 = \tau(z_1), \quad T(\tau(z_2)) + z_2 = \tau(z_2).$$

It is easy to see that

$$\begin{aligned} ||\tau(z_1) - \tau(z_2)|| &\le ||z_1 - z_2|| + ||T(\tau(z_1)) - T(\tau(z_2))|| \\ &\le ||z_1 - z_2|| + \alpha||\tau(z_1) - \tau(z_2)||, \end{aligned}$$

which implies that

$$||\tau(z_1) - \tau(z_2)|| \le \frac{1}{1-\alpha}||z_1 - z_2||.$$

Hence, $\tau : S(D) \to D$ is continuous. Since S is completely continuous on D, $\tau S : D \to D$ is completely continuous. By Leray–Schauder Fixed Point Theorem, there exists $x_0 \in D$ such that $\tau S(x_0) = x_0$. Since

$$T(\tau(S(x_0))) + S(x_0) = \tau(S(x_0)),$$

we obtain $Tx_0 + Sx_0 = x_0$. This is the end.

2.5.2 Fixed Point Theorem of Expansion and Compression in Norm Form

First, we prove a lemma.

Lemma 2.12. *Suppose that X is an infinite dimensional real Banach space, $\Omega \subset X$ is a bounded open subset, and $F : \bar{\Omega} \to X$ is completely continuous. Assume the following two conditions hold:*

(1) $\inf_{x\in\partial\Omega} ||F(x)|| > 0$.
(2) $F(x) = \mu x,\ x \in \partial\Omega \Longrightarrow \mu \notin (0, 1]$.

Then $\deg(id - F, \Omega, 0) = 0$.

Proof. First, we claim that

$$\inf_{x\in\partial\Omega, 0\leq\mu\leq 1} ||\mu x - F(x)|| =: \alpha > 0.$$

We assume, by contradiction, that there exist $x_n \in \partial\Omega$, $\mu_n \in [0, 1]$, such that $||\mu_n x_n - F(x_n)|| \to 0$ $(n \to \infty)$. Since F is completely continuous and $\{\mu_n\}$ is bounded, there exists a subsequence such that $\mu_{n_k} \to \mu_0$, $F(x_{n_k}) \to y_0$. Clearly, $\mu_0 > 0$. Then, $x_{n_k} \to \frac{1}{\mu_0} y_0 =: x_0 \in \partial\Omega$. So, $\mu_0 x_0 = F(x_0)$, $0 < \mu_0 \leq 1$, which is a contradiction.

Take a finite dimensional subspace X_n of X and a continuous bounded mapping $F_n : \bar{\Omega} \to X_n$, such that

$$\sup_{x\in\bar{\Omega}} ||F(x) - F_n(x)|| < \alpha.$$

Since X is an infinite dimensional Banach space, we can take $y_{n+1} \in X \backslash X_n$ with $||y_{n+1}|| = 1$. Denote by X_{n+1} the subspace spanned by X_n and y_{n+1}. We can view $F_n : \bar{\Omega} \to X_{n+1}$. By the definition of Leray–Schauder degree, we have

$$\deg(id - F, \Omega, 0) = \deg(id - F_n, \Omega_{n+1}, 0),$$

where $\Omega_{n+1} = X_{n+1} \cap \Omega$. By the conditions and the above claim, we arrive at that whenever $x \in \partial\Omega_{n+1}$ and $0 \leq \mu \leq 1$,

$$||\mu x - F_n(x)|| \geq ||\mu x - F(x)|| - ||F(x) - F_n(x)|| > 0.$$

So, $id - F_n$ and $-F_n$ are homotopy on Ω_{n+1}. Hence,

$$\deg(id - F_n, \Omega_{n+1}, 0) = \deg(-F_n, \Omega_{n+1}, 0).$$

By the dimensionality reduction property of Brouwer degree, $\deg(-F_n, \Omega_{n+1}, 0) = 0$. This is the end.

Corollary 2.6. *Let Ω be a bounded open subset of an infinite dimensional real Banach space X, $0 \notin \partial\Omega$. Suppose that $F : \bar{\Omega} \to X$ is completely continuous and satisfies*

$$||F(x)|| \geq ||x||, \; F(x) \neq x, \quad \forall x \in \partial\Omega.$$

Then $\deg(id - F, \Omega, 0) = 0$.

Proof. It is clear that the condition (1) in Lemma 2.12 holds. Next, we show that the condition (2) in Lemma 2.12 also holds. In fact, we assume, by contradiction, that there exist $x_0 \in \partial\Omega$ and $0 < \mu_0 \leq 1$, such that $F(x_0) = \mu_0 x_0$. Then

$$\mu_0 \neq 1, \; x_0 \neq 0, \; ||F(x_0)|| = \mu_0||x_0|| < ||x_0||,$$

which is a contradiction. This is the end.

Theorem 2.29 (Fixed Point Theorem of Expansion and Compression in Norm Form). *Let Ω_1 and Ω_2 be two bounded open subsets of an infinite dimensional real Banach space, $0 \in \Omega_1, \bar{\Omega}_1 \subset \Omega_2$, and $F : \bar{\Omega}_2 \backslash \Omega_1 \to X$ be completely continuous. Suppose that one of the following condition holds:*

(H_1) $x \in \partial\Omega_1 \Rightarrow ||F(x)|| \leq ||x||$, $x \in \partial\Omega_2 \Rightarrow ||F(x)|| \geq ||x||$ (*Expansion in Norm Form*).

(H_2) $x \in \partial\Omega_1 \Rightarrow ||F(x)|| \geq ||x||$, $x \in \partial\Omega_2 \Rightarrow ||F(x)|| \leq ||x||$ (*Compression in Norm Form*).

Then F admits at least one fixed point in $\bar{\Omega}_2 \backslash \Omega_1$.

Proof. According to the extension theorem, F can be extended into a completely continuous mapping from $\bar{\Omega}_2 \to X$. We can suppose that $F(x) \neq x$, $\forall x \in \partial\Omega_1 \cup \partial\Omega_2$, otherwise the conclusion of theorem is valid.

Assume that the condition (H_1) holds. By Corollaries 2.6 and 2.5, we obtain

$$\deg(id - F, \Omega_2, 0) = 0, \ \deg(id - F, \Omega_1, 0) = 1.$$

By Additivity, we arrive at

$$\deg(id - F, \Omega_2 \backslash \bar{\Omega}_1, 0) = \deg(id - F, \Omega_2, 0) - \deg(id - F, \Omega_1, 0) = -1.$$

By Kronecker existence theorem, we know that F admits at least one fixed point in $\Omega_2 \backslash \bar{\Omega}_1$.

Assume that the condition (H_2) holds. The similar reason implies

$$\deg(id - F, \Omega_2 \backslash \bar{\Omega}_1, 0) = \deg(id - F, \Omega_2, 0) - \deg(id - F, \Omega_1, 0) = 1.$$

By Kronecker existence theorem, F admits at least one fixed point in $\Omega_2 \backslash \bar{\Omega}_1$. This is the end.

Remark. For finite dimensional space X, Theorem 2.29 does not hold. For example, consider $\Omega_1 = \{(x, y);\ x^2 + y^2 < r_1^2\}$, $\Omega_2 = \{(x, y);\ x^2 + y^2 < r_2^2\}$ $(0 < r_1 < r_2)$ in $\mathbb{R}^2$. We use the complex number $z = x + iy = re^{i\theta}$ to represent the point (x, y) in $\mathbb{R}^2$. Let $F(z) = z_1$, $z_1 = re^{i(\theta + \frac{\pi}{4})}$. Clearly, $F : \mathbb{R}^2 \to \mathbb{R}^2$ is bounded and continuous (completely continuous), $||F(z)|| = r_1 = ||z||$ holds on $\partial\Omega_1$, and $||F(z)|| = r_2 = ||z||$ holds on $\partial\Omega_2$. But, it is clear that F does not have any fixed point in $\bar{\Omega}_2 \backslash \Omega_1$.

2.5.3 Borsuk Theorem

Theorem 2.30 (Borsuk). *Let $\Omega \subset X$ be a bounded open set with $0 \in \Omega$ and symmetric with respect to the origin. Suppose that $f = id - F : \bar{\Omega} \to X$ is a completely continuous field satisfying that $f(x) \neq tf(-x)$, whenever $x \in \partial\Omega$, $t \in [0, 1]$. Then* $\deg(f, \Omega, 0)$ *is odd number, so that F admits at least one fixed point in Ω.*

Proof. Let

$$H(x, t) = \frac{1}{1+t} F(x) - \frac{t}{1+t} F(-x),$$
$$h_t(x) = x - H(x, t) = \frac{1}{1+t} f(x) - \frac{t}{1+t} f(-x).$$

Then $H : \bar{\Omega} \times [0, 1] \to X$ is compact continuous. By the assumption, $h_t(x) \neq 0$ whenever $x \in \partial\Omega$ and $t \in [0, 1]$; $h_0 = f$ as $t = 0$; $h_1(x) = g(x) = x - G(x) = \frac{1}{2}f(x) - \frac{1}{2}f(-x)$ as $t = 1$. By compact homotopy invariance, we obtain

$$\deg(f, \Omega, 0) = \deg(g, \Omega, 0).$$

Since g is odd mapping on Ω, G is odd on Ω. By Theorem 2.17, there exist a finite dimensional subspace X_n and a continuous mapping $G_n : \bar{\Omega} \to X_n$ such that

$$\sup_{x\in\bar{\Omega}} ||G(x) - G_n(x)|| < \epsilon = \text{dist}(0, g(\partial\Omega)) > 0.$$

Let

$$\tilde{G}_n(x) = \frac{1}{2}(G_n(x) - G_n(-x)).$$

Then $\tilde{G}_n$ is odd mapping, and

$$\sup_{x\in\bar{\Omega}} ||G(x) - \tilde{G}_n(x)|| \leq \frac{1}{2} \sup_{x\in\bar{\Omega}} ||G(x) - G_n(x)|| + \frac{1}{2} \sup_{x\in\bar{\Omega}} ||G(-x) - G_n(-x)|| < \epsilon.$$

Denote

$$g_n(x) = x - \tilde{G}_n(x).$$

By the definition of Leray–Schauder degree and Borsuk Theorem, we obtain

$$\deg(g, \Omega, 0) = \deg(g_n, \Omega \cap X_n, 0) = \text{odd number}.$$

Therefore, $\deg(f, \Omega, 0) =$ odd number. This is the end.

Corollary 2.7 (Borsuk). *Let $\Omega \subset X$ be a bounded open set containing the origin and symmetric with respect to the origin. Suppose that $F : \bar{\Omega} \to X$ is completely continuous and satisfies $F(-x) = -F(x)$ whenever $x \in \partial\Omega = S$. Then F admits at least one fixed point in $\bar{\Omega}$.*

Proof. If F has a fixed point on S, then the conclusion naturally is valid. So, let us assume that F has no fixed point on S. Let $f = id - F$. Then, f is a completely continuous field, $f(-x) = -f(x)$ and $f(x) \neq 0$ whenever $x \in S$. Hence, $tf(-x) = -tf(x) \neq f(x)$ whenever $x \in S$ and $t \in [0, 1]$. By Theorem 2.30, F admits at least one fixed point in Ω. This is the end.

Similarly, we can also prove the following theorem.

Theorem 2.31. *Suppose that $\Omega \subset X$ is an open subset, $f = id - F : \Omega \to X$ is a completely continuous field and locally one-to-one. Then f is an open mapping.*

2.6 Topological Degree of Cone Mappings

In this section we always assume that K is a closed cone in Banach space X. By Dugundji Theorem [3] (an extension version of Tietze Theorem), K is a retract of X, that is, there is a continuous mapping $r : X \to K$ such that $r(x) = x$ whenever $x \in K$.

Definition 2.6.1. Let K be a closed cone in Banach space X, $D \subset X$. If $F(D) \subset K$, the mapping $F : D \to K$ is said to be a **cone mapping**.

In the sequel, we suppose that $D \subset K$ is a bounded open subset in K (refer to a relatively open subset, where the metric of K is induced by the norm of X), $F : D \to K$ is a cone mapping and completely continuous. Let $f = id - F$. Suppose $0 \notin f(\partial D)$, where ∂D is the boundary of D relative to K. Take R sufficiently large, such that $D \subset B_R(0)$ (a ball with radius R centered at point 0). For a cone mapping, we define the following topological degree.

Definition 2.6.2.

$$\deg_K(id - F, D, 0) = \deg(id - F \circ r, B_R(0) \cap r^{-1}(D), 0).$$

When the meaning is clear, it can be abbreviated as $\deg(f, D, 0)$.

We also use the notation

$$i(F, D, K) =: \deg_K(id - F, D, 0),$$

that is called the **fixed point index** of F over D with respect to K. Clearly, $F \circ r : B_R(0) \cap r^{-1}(D) \to K$ is completely continuous. To verify the validity of the definition, it is necessary to first explain $0 \notin (id - F \circ r)(\partial(B_R(0) \cap r^{-1}(D)))$. In fact, assume, by contradiction, that there is an $x \in \partial(B_R(0) \cap r^{-1}(D))$ such that $x = F \circ r(x)$. Then $x \in K$. Hence, $x = F(x)$, and $x \in K \cap \partial(B_R(0) \cap r^{-1}(D))$. This will imply $x \in \partial D$, which is a contradiction. Therefore, the Leray–Schauder degree

$$\deg(id - F \circ r, B_R(0) \cap r^{-1}(D), 0) \tag{2.23}$$

is well defined. Secondly, we also need to clarify that the value in (2.23) is independent of the choice of r, R.

Lemma 2.13. *The value in* (2.23) *is independent of the choice of* r, R.

Proof. (1) First, we show that the value in (2.23) is independent of R. Suppose that there is another real number R_1, such that $D \subset B_{R_1}(0)$. Without loss of generality, assume $R_1 > R$. Clearly, the fixed point of $F \circ r$ belongs to D, and $D \subset B_R(0) \cap r^{-1}(D) \subset B_{R_1}(0) \cap r^{-1}(D)$. Thus, $F \circ r$ has no fixed point in $\overline{B_{R_1}(0) \cap r^{-1}(D)} \backslash (B_R(0) \cap r^{-1}(D))$. By the excision of Leray–Schauder degree,

$$\deg(id - F \circ r, B_{R_1}(0) \cap r^{-1}(D), 0) = \deg(id - F \circ r, B_R(0) \cap r^{-1}(D), 0).$$

(2) We prove that the value in (2.23) is independent of the choice of retraction. Suppose that $r_1 : X \to K$ is another retraction. We will prove

$$\begin{aligned} &\deg(id - F \circ r_1, B_R(0) \cap r_1^{-1}(D), 0) \\ &\quad = \deg(id - F \circ r, B_R(0) \cap r^{-1}(D), 0). \end{aligned} \tag{2.24}$$

Let $U = B_R(0) \cap r^{-1}(D) \cap r_1^{-1}(D)$. Then U is an open subset of X, and $D \subset U$. Thus, $F \circ r$ has no fixed point in $\overline{B_R(0) \cap r^{-1}(D)} \backslash U$, and $F \circ r_1$ has no fixed point in $\overline{B_R(0) \cap r_1^{-1}(D)} \backslash U$. Hence,

$$\begin{aligned} \deg(id - F \circ r, B_R(0) \cap r^{-1}(D), 0) &= \deg(id - F \circ r, U, 0), \\ \deg(id - F \circ r_1, B_R(0) \cap r_1^{-1}(D), 0) &= \deg(id - F \circ r_1, U, 0). \end{aligned} \tag{2.25}$$

Denote

$$H(x,t) = r(tF(r(x)) + (1-t)F(r_1(x))).$$

Clearly, $H : \bar{U} \times [0,1] \to X$ is completely continuous. We claim that $H(x,t) \neq x$ whenever $(x,t) \in \partial U \times [0,1]$. In fact, assume, by contradiction, that there would be $(x_0, t_0) \in \partial U \times [0,1]$, such that $H(x_0, t_0) = x_0$, that is, $r(tF(r(x_0)) + (1-t)F(r_1(x_0))) = x_0$. Then, $x_0 \in K$, $r(x_0) = x_0$, $r_1(x_0) = x_0$, $x_0 = r(t_0F(x_0) + (1-t_0)F(x_0)) = r(F(x_0)) = F(x_0)$. Hence, $x_0 \in D \subset U$, which is in contradiction with $x_0 \in \partial U$.

Since $U \subset \overline{r^{-1}(D)} \cap \overline{r_1^{-1}(D)} \subset r^{-1}(\bar{D}) \cap r_1^{-1}(\bar{D})$, it follows that $r(x) \in \bar{D}$, $r_1(x) \in \bar{D}$ whenever $x \in U$. Thus,

$$H(x,0) = r(F(r_1(x))) = F(r_1(x)), \quad H(x,1) = r(F(r(x))) = F(r(x)).$$

By the homotopy invariance of Leray–Schauder degree, we arrive at

$$\deg(id - F \circ r_1, U, 0) = \deg(id - F \circ r, U, 0).$$

By (2.25), (2.24) holds. This is the end.

The rationality of defining the topological degree of cone mapping has been explained above. Below are some of its properties and applications.

Theorem 2.32. *The topological degree* $\deg(f, D, 0)$ *of cone mapping satisfies the following fundamental properties.*

(i) **(Normality)** *Suppose* $x_0 \in D$. *Then* $\deg(id - x_0, D, 0) = 1$.

(ii) **(Additivity)** *Let* D_1, D_2 *be two relative open subsets of* D *and disjoint. Suppose that* $F : \bar{D} \to K$ *is completely continuous,* $f = id - F$, *and* $0 \notin f(\bar{D} \backslash (D_1 \cup D_2))$. *Then*

$$\deg(f, D, 0) = \deg(f, D_1, 0) + \deg(f, D_2, 0).$$

(iii) **(Compact homotopy invariance)** *Suppose that* $H : \bar{D} \times [0,1] \to K$ *is compact continuous, and* $h_t(x) = x - H(x,t) \neq 0$ *whenever* $x \in \partial D$ *and* $t \in [0,1]$. *Then* $\deg(h_t, D, 0)$ *is independent of* t.

Proof. (i) Suppose that $F : \bar{D} \to K$ is a constant mapping, that is, $F(x) = x_0$. Thus, if $x \in \overline{B_R(0) \cap r^{-1}(D)}$, $(id - F \circ r)(x) = x - x_0$ and $x_0 \in D \subset B_R(0) \cap r^{-1}(D)$. By the properties of Leray–Schauder degree, it follows that

$$\begin{aligned} &\deg\left(id - F \circ r, B_R(0) \bigcap r^{-1}(D), 0\right) \\ &= \deg\left(id - x_0, B_R(0) \bigcap r^{-1}(D), 0\right) \\ &= \deg\left(id, B_R(0) \bigcap r^{-1}(D), x_0\right) \\ &= 1. \end{aligned}$$

(ii) Clearly, $F \circ r$ has no fixed point in $\overline{B_R(0) \cap r^{-1}(D)} \setminus ((B_R(0) \cap r^{-1}(D_1)) \cup (B_R(0) \cap r^{-1}(D_2)))$, and $B_R(0) \cap r^{-1}(D_1)$ and $B_R(0) \cap r^{-1}(D_2)$ are disjoint. By the additivity of Leray–Schauder degree, we arrive at

$$\begin{aligned} &\deg\left(id - F \circ r, B_R(0) \bigcap r^{-1}(D), 0\right) \\ &= \deg\left(id - F \circ r, B_R(0) \bigcap r^{-1}(D_1), 0\right) \\ &\quad + \deg\left(id - F \circ r, B_R(0) \bigcap r^{-1}(D_2), 0\right). \end{aligned}$$

(iii) Consider $H(r(x), t) : \overline{(B_R(0) \cap r^{-1}(D))} \times [0, 1] \to K$, which is completely continuous. We claim that $H(r(x), t) \neq x$ whenever $x \in \partial(B_R(0) \cap r^{-1}(D))$ and $t \in [0, 1]$.

In fact, assume, by contradiction, that there are $x_0 \in \partial(B_R(0) \cap r^{-1}(D))$ and $t_0 \in [0, 1]$, such that $H(r(x_0), t_0) = x_0$. Then, $r(x_0) \in \bar{D}$, $x_0 = H(r(x_0), t_0) \in K$. Thus, $r(x_0) = x_0$, $H(x_0, t_0) = x_0$. At this moment, we can know again $x_0 \in \partial D$, which is a contradiction.

By the homotopy invariance of Leray–Schauder degree, we obtain that

$$\deg\left(id - H(r(\cdot), t), B_R(0) \bigcap r^{-1}(D), 0\right)$$

is independent of t. This is the end.

Theorem 2.33. *Let D be a bounded open subset in K, $F : \bar{D} \to K$ be compact continuous, and $f = id - F$. Suppose that there is an $x_0 \in D$ such that, whenever $x \in \partial D$ and $\lambda \geq 1$,*

$$F(x) - x_0 \neq \lambda(x - x_0).$$

Then $\deg(f, D, 0) = 1$.

Proof. Let $h_t(x) = x - [x_0 + t(F(x) - x_0)]$, $x \in \bar{D}$, $t \in [0, 1]$. Then, $h_t(x) \neq 0$, whenever $x \in \partial D$ and $t \in [0, 1]$. By compact homotopy invariance and normality, we arrive at

$$\deg(f, D, 0) = \deg(h_1, D, 0) = \deg(h_0, D, 0) = 1.$$

This is the end.

Corollary 2.8. *Let D be a bounded open subset in K, $0 \in D$. Suppose that $F : \bar{D} \to K$ is completely continuous, and satisfies*

$$F(x) \not\geq x, \quad x \in \partial D. \tag{2.26}$$

Then $\deg(id - F, D, 0) = 1$.

Proof. By (2.26), $F(x) \neq \lambda x$, whenever $x \in \partial D$ and $\lambda \geq 1$. Letting $x_0 = 0$, Theorem 2.33 implies

$$\deg(id - F, D, 0) = 1.$$

This is the end.

Theorem 2.34. *Let D be a bounded open subset in K, and $F : \bar{D} \to K$ be completely continuous. Suppose that there is a $p \in K \backslash \{0\}$ such that $x - F(x) \neq \lambda p$, whenever $x \in \partial D$ and $\lambda \geq 0$. Then*

$$\deg(id - F, D, 0) = 0.$$

Proof. Take λ_0 such that $\lambda_0 > ||p||^{-1} \sup_{x \in \bar{D}} ||x - F(x)||$. Let $h_t(x) = x - F(x) - t\lambda_0 p$. Then, $F(x) + t\lambda_0 p \in K$ whenever $x \in \bar{D}$ and $t \geq 0$; $h_t(x) \neq 0$ whenever $x \in \partial D$ and $0 \leq t \leq 1$. Thus, $\deg(h_t, D, 0)$ is well defined. By compact homotopy invariance,

$$\deg(id - F, D, 0) = \deg(h_0, D, 0) = \deg(h_1, D, 0).$$

The choice of λ_0 implies that for any $x \in \bar{D}$,

$$x - F(x) \neq \lambda_0 p.$$

Therefore, $\deg(h_1, D, 0) = 0$. This is the end.

Corollary 2.9. *Let D be a bounded open subset in K, and $F : \bar{D} \to K$ be completely continuous. Suppose*

$$F(x) \nleq x, \quad x \in \partial D. \tag{2.27}$$

Then $\deg(id - F, D, 0) = 0$.

Proof. Letting $p \in K \backslash \{0\}$, then (2.27) implies that whenever $x \in \partial D$,

$$x - F(x) \neq tp, \quad t \geq 0.$$

By Theorem 2.34, $\deg(id - F, D, 0) = 0$. This is the end.

Theorem 2.35. *Let D_1, D_2 be two bounded open subsets in K, $\bar{D}_1 \subset D_2$, and $0 \in D_1$. Suppose that $F : \bar{D}_2 \to K$ is completely continuous and one of the following two conditions holds:*

(i) *cone compression:*

$$F(x) \nleq x, \quad x \in \partial D_1,$$
$$F(x) \ngeq x, \quad x \in \partial D_2.$$

(ii) *cone expansion:*

$$F(x) \ngeq x, \quad x \in \partial D_1,$$
$$F(x) \nleq x, \quad x \in \partial D_2.$$

Then F admits at least one fixed point in $\bar{D}_2 \backslash D_1$.

Proof. Suppose that F has no fixed point on $\partial D_1 \cup \partial D_2$ without loss of generality. Let $f = id - F$. Then, $\deg(f, D_i, 0)$, $i = 1, 2$ and $\deg(f, D_2 \backslash \bar{D}_1, 0)$ are well defined, and

$$\deg(f, D_2 \backslash \bar{D}_1, 0) = \deg(f, D_2, 0) - \deg(f, D_1, 0).$$

If the condition (i) holds, Corollaries 2.8 and 2.9 imply

$$\deg(f, D_2, 0) = 1, \quad \deg(f, D_1, 0) = 0,$$

respectively. When condition (ii) is met, the same reason can imply

$$\deg(f, D_2, 0) = 0, \quad \deg(f, D_1, 0) = 1.$$

Hence,

$$\deg(f, D_2 \backslash \bar{D}_1, 0) \neq 0.$$

Therefore, f admits at least one fixed point $D_2 \backslash \bar{D}_1$. This is the end.

Lemma 2.14. *Let K be a closed cone in Banach space X, C be a compact set in K, and $0 \notin C$. Then, $0 \notin \overline{\text{conv}}\, C$ (the closed convex hull of C).*

Proof. According to the conclusion (iii) of Theorem 1.44, for every $x \in C$, there exists $f_x \in K^*$, such that $f_x(x) \geq 1$. Thus, there exists a small neighborhood $B(x, \epsilon_x)$ of x, such that, whenever $y \in B(x, \epsilon_x)$, $f_x(y) \geq \frac{1}{2}$. Since C is compact and $\{B(x, \epsilon_x);\ x \in C\}$ is an open cover of C, there exists a finite number of $B(x_i, \epsilon_{x_i})$, $i = 1, 2, \ldots, n$, such that $C \subset \bigcup_{i=1}^n B(x_i, \epsilon_{x_i})$. Let $f = \sum_{i=1}^n f_{x_i}$. Then, $f \in K^*$, and, for every $x \in C$, there exists some i, $1 \leq i \leq n$, such that $x \in B(x_i, \epsilon_{x_i})$, which implies that

$$f(x) = \sum_{j=1}^n f_{x_j}(x) \geq f_{x_i}(x) \geq \frac{1}{2}.$$

Hence, for each convex combination of any finite number of elements $y_1, \ldots, y_m$ in C,

$$y = \sum_{i=1}^m \alpha_i y_i, \quad 0 \leq \alpha_i \leq 1, \ \sum_{i=1}^m \alpha_i = 1,$$

we have

$$f(y) = \sum_{i=1}^m \alpha_i f(y_i) \geq \frac{1}{2}.$$

Therefore, for any $y \in \overline{\text{conv}}\, C$, $f(y) \geq \frac{1}{2}$, which implies $0 \notin \overline{\text{conv}}\, C$. This is the end.

Theorem 2.36. *Let D_1, D_2 be two bounded open subsets in K, $\bar{D}_1 \subset D_2$, and $0 \in D_1$. Suppose that $F : \bar{D}_2 \to K$ is completely continuous and satisfies the following conditions:*

(i) *whenever $\lambda > 1$ and $x \in \partial D_2$, $F(x) \neq \lambda x$,*
(ii) *whenever $0 < \lambda < 1$ and $x \in \partial D_1$, $F(x) \neq \lambda x$,*
(iii) $\inf_{x \in \partial D_1} ||F(x)|| > 0$.

Then F admits at least one fixed point in $\bar{D}_2 \backslash D_1$.

Proof. Suppose that F has no fixed point on $\partial D_1 \cup \partial D_2$, otherwise the conclusion holds. By the condition (i) and Theorem 2.33, we know $\deg(id - F, D_2, 0) = 1$.

On the other hand, by the Dugundji extension theorem, there exists a completely continuous mapping $F_1 : \bar{D}_1 \to \overline{\text{conv}}\{F(x);\ x \in \partial D_1\}$, such that, whenever $x \in \partial D_1$, $F_1(x) = F(x)$. By the condition (iii) and Lemma 2.14, we obtain

$$\inf_{x \in \bar{D}_1} ||F_1(x)|| = \alpha > 0.$$

Let $\rho = \sup_{x \in \bar{D}_1} ||x||$ and take $s > \max\{1, \frac{\rho}{\alpha}\}$. When $x \in \partial D_1$ and $t \in [0, 1]$, it follows from the condition (ii) that

$$(1 - t)F(x) + tsF_1(x) \neq x.$$

By the compact homotopy invariance, we have

$$\deg(id - F, D_1, 0) = \deg(id - sF_1, D_1, 0).$$

Since $x = sF_1(x)$ implies $s \leq \frac{\rho}{\alpha}$, we obtain

$$\deg(id - sF_1, D_1, 0) = 0.$$

By Additivity,

$$\deg(id - F, D_2 \backslash \bar{D}_1, 0) = \deg(id - F, D_2, 0) - \deg(id - F, D_1, 0) = 1.$$

This is the end.

There have been many results on the existence of multiple positive solutions, but we only give two theorems as follows. For more information, please refer to other literature [13, 22].

Theorem 2.37. *Let D_i, $i = 1, 2, \ldots, n$ be bounded open subsets in K, $0 \in D_1$, and $\bar{D}_i \subset D_{i+1}$, $i = 1, \ldots, n-1$. Suppose that $F : \bar{D}_n \to K$ is completely continuous and satisfies the following conditions:*

(1) $F(x) \ngeq x, \quad x \in \partial D_{2i-1},\ 1 \leq 2i - 1 \leq n,$
(2) $F(x) \nleq x, \quad x \in \partial D_{2i},\ 1 \leq 2i \leq n.$

Then F has at least n fixed points $x_1, \ldots, x_n$, where $x_i \in D_i \backslash \bar{D}_{i-1}$, $D_0 = \emptyset$.

Proof. Corollaries 2.8 and 2.9 imply

$$\deg(id - F, D_i \backslash \bar{D}_{i-1}, 0) \neq 0.$$

Thus, F has at least one fixed point in $D_i \backslash \bar{D}_{i-1}$. This is the end.

Theorem 2.38. *Let $D_0, D_1, \ldots, D_n$ $(n \geq 2)$ be bounded open subsets in K, $D_i \subset D_0$, $1 \leq i \leq n$, and $D_i \cap D_j = \emptyset$ as $i \neq j$, $1 \leq i, j \leq n$. Suppose that $F : \bar{D} \to K$ is completely continuous and satisfies the following conditions:*

(1) *there is an $x_0 \in D_0$ such that, whenever $\lambda \geq 1$ and $x \in \partial D_0$,*

$$F(x) - x_0 \neq \lambda(x - x_0),$$

(2) *there exist $x_i \in D_i$, $i = 1, 2, \ldots, n$ such that, whenever $\lambda \geq 1$ and $x \in \partial D_i$,*

$$F(x) - x_i \neq \lambda(x - x_i).$$

Then F has at least $n+1$ fixed points y_i, $i = 0, 1, \ldots, n$ in D_0, where $y_i \in D_i$, $y_0 \in \bar{D}_0 \backslash \bigcup_{i=1}^n \bar{D}_i$.

Proof. By Theorem 2.33, $\deg(id - F, D_i, 0) = 1$, for each i, $i = 0, 1, 2, \ldots, n$. Thus, F has at least one fixed point y_i in D_i. Suppose that F has no fixed point on ∂D_0 without loss of generality. Then $\deg(id - F, D_0, 0)$ and $\deg(id - F, D_0 \backslash (\bigcup_{i=1}^n \bar{D}_i), 0)$ are well defined. By Additivity,

$$\begin{aligned} &\deg\left(id - F, D_0 \Bigg\backslash \left(\bigcup_{i=1}^n \bar{D}_i\right), 0\right) \\ &\quad = \deg(id - F, D_0, 0) - \sum_{i=1}^n \deg(id - F, D_i, 0) = 1 - n. \end{aligned}$$

Hence, F has at least one fixed point y_0 in $D_0 \backslash (\bigcup_{i=1}^n \bar{D}_i)$. This is the end.

2.7 The Coincidence Degree

The coincidence degree theory mainly aims to provide an approach to prove the existence of solution for the following equation of form

$$Lu = N(u),$$

where $L : \text{dom}L \subset X \to Z$ is the Fredholm operator, that is, the range $\text{Im}\, L$ of L is a closed space of Z; the kernel $\text{Ker}\, L = L^{-1}(0)$ of L is finite dimensional space; $\text{codim}\,\text{Im}\, L := \dim(Z/\text{Im}\, L) < \infty$ (the codimension of the range of L is finite). We say

$$\text{ind}\,(L) := \dim \text{Ker}\, L - \text{codim}\,\text{Im}\, L$$

to be the **index** of L.

The main idea is to transform the solving $Lu = N(u)$ into solving $u = Mu$, where M is a completely continuous operator.

When L is the Fredholm operator, there is the following decompositions:

$$X = \text{Ker}\, L \oplus X_1,$$
$$Z = \text{Im}\, L \oplus Z_1,$$

and there exist projection operators $P : X \to \text{Ker}\, L$, $Q : Z \to Z_1$, such that

$$X = \text{Ker}\, L \oplus X_1 = \text{Im}\, P \oplus \text{Ker}\, P, \quad \text{Im}\, P = \text{Ker}\, L,$$
$$Z = Z_1 \oplus \text{Im}\, L = \text{Im}\, Q \oplus \text{Ker}\, Q, \quad \text{Ker}\, Q = \text{Im}\, L.$$

Clearly, $\text{Ker}\, P = \text{Im}(I - P)$, $\text{Ker}\, Q = \text{Im}(I - Q)$, $\text{ker}L \cap (\text{dom } L \cap X_1) = \{0\}$. Thus,

$$L\big|_{\text{dom}L\cap\text{Ker}P} : \text{dom } L \cap X_1 \subset (I - P)X \to \text{Im}\, L$$

is an invertible mapping, and its inverse mapping is denoted as K_P.

Suppose that L is the Fredholm operator with index 0. Then

$$\dim \text{Ker}\, L = \dim(Z/\text{Im}\, L).$$

So, it can be inferred that there exists an isomorphic mapping

$$J : \text{Im}\, Q \simeq \text{Ker}\, L.$$

For u satisfying $Lu = N(u)$, we have

$$L(Pu + (I - P)u) = QN(u) + (I - Q)N(u).$$

Since $\operatorname{Im} P = \operatorname{Ker} L$ and $\operatorname{Ker} Q = \operatorname{Im} L$, it follows that $LPu = 0$, $QN(u) = 0$. Thus,

$$L(I - P)u = (I - Q)N(u).$$

Hence,

$$(I - P)u = K_P(I - Q)N(u),$$

which implies that

$$u = Pu + K_P(I - Q)N(u).$$

Therefore, it can be written as

$$u = Pu + JQN(u) + K_P(I - Q)N(u).$$

On the other hand, if $u = Pu + JQN(u) + K_P(I - Q)N(u)$, then $(I - P)u = JQN(u) + K_P(I - Q)N(u)$. Thus,

$$L(I - P)u = LJQN(u) + LK_P(I - Q)N(u).$$

Hence,

$$L(I - P)u = (I - Q)N(u).$$

Since $JQN(u) \in \operatorname{Ker} L$ and $K_P(I - Q)N(u) \in \operatorname{dom} L \cap \operatorname{Ker} P$, then $JQN(u) = 0$, that is, $QN(u) = 0$. So, we obtain that $Lu = N(u)$.

As explained above, the problem of solving $Lu = N(u)$ is transformed into a problem of solving

$$u = Pu + JQN(u) + K_P(I - Q)N(u) =: Mu.$$

Definition 2.7.1. Let $N : X \to Z$ be continuous, and $\Omega \subset X$ be a bounded open subset. Suppose that $QN(\bar{\Omega})$ is bounded, and $K_P(I - Q)N : \bar{\Omega} \to X$ is compact. Then N is said to be ***L*-compact** on $\bar{\Omega}$.

Definition 2.7.2 (Coincidence Degree). Let $\Omega \subset X$ be a bounded open subset. Suppose that $0 \notin (L-N)(\operatorname{dom} L \cap \partial\Omega)$. Then the **coincidence degree** of L and N on Ω is defined by

$$\mathrm{d_C}[(L,N),\Omega] := \deg_{\mathrm{LS}}(I-M,\Omega,0),$$

where $M = P + JQN + K_P(I-Q)N$ and $\deg_{\mathrm{LS}}$ is the Leray–Schauder degree.

Remark. It can be shown that the definition of degree above is independent of projection P, Q, and algebraic isomorphism J.

By utilizing the properties of Leray–Schauder degree, it is easy to write some properties of coincidence degree.

Theorem 2.39 (Homotopy Invariance). *Let L be the Fredholm operator with index 0. Suppose that $\Phi : \bar{\Omega} \to Z$ is continuous, such that the mappings $Q\Phi$ and $K_P(I-Q)\Phi$ are compact continuous on $\bar{\Omega}$, and satisfy the following conditions:*

(1) $(1-\lambda)(L-N)u + \lambda(L-\Phi)u \neq 0, \quad \forall u \in \operatorname{dom} L \cap \partial\Omega, \lambda \in [0,1]$,
(2) $\mathrm{d_C}[(L,\Phi),\Omega] \neq 0$.

Then the equation

$$Lu = N(u)$$

has at least one solution in the region $\operatorname{dom} L \cap \bar{\Omega}$.

Theorem 2.40 (Mawhin). *Let L be the Fredholm operator with index 0, and N be L-compact on $\bar{\Omega}$. Suppose that the following conditions hold:*

(1) $Lx \neq \lambda N(x), \ \forall x \in \operatorname{dom} L \cap \partial\Omega, \ \lambda \in (0,1)$.
(2) *For any* $x \in \operatorname{Ker} L \cap \partial\Omega$, $QN(x) \neq 0$, *and*

$$\deg(JQN, \Omega \cap \operatorname{Ker} L, 0) \neq 0.$$

Then $Lx = N(x)$ has at least one solution in $\operatorname{dom} L \cap \bar{\Omega}$.

Proof. Assume that the equation $Lx = N(x)$ has no solution on $\operatorname{dom} L \cap \partial\Omega$, otherwise the conclusion holds. From the previous explanation, it is known that

$$Lx = \lambda N(x) \Longleftrightarrow x = Px + JQN(x) + \lambda K_P(I - Q)N(x).$$

We define the mapping $H : \bar{\Omega} \times [0, 1] \to X$ as follows:

$$H(x, \lambda) = Px + JQN(x) + \lambda K_P(I - Q)N(x).$$

Then $H : \bar{\Omega} \times [0, 1] \to X$ is a completely continuous mapping. We claim that

$$x \neq H(x, \lambda), \quad \forall x \in \partial\Omega,\ \lambda \in [0, 1].$$

In fact, if it is false, then there exist $x_0 \in \partial\Omega$ and $\lambda_0 \in [0, 1]$, such that

$$x_0 = H(x_0, \lambda_0) = Px_0 + JQN(x_0) + \lambda_0 K_P(I - Q)N(x_0).$$

It is easy to see that $x_0 \in \operatorname{dom} L$. Thus, $x_0 \in \operatorname{dom} L \cap \partial\Omega$. By the condition (1) and assumption, we have $\lambda_0 = 0$. So,

$$x_0 = Px_0 + JQN(x_0).$$

Hence, $JQN(x_0) = 0$, that is, $QN(x_0) = 0$. This is a contradiction with the condition (2). Therefore, the claim holds.

By the definition of Leray–Schauder degree and compact homotopy invariance, we obtain that

$$\begin{aligned}
\mathrm{d_C}[(L, N), \Omega] &= \deg_{\mathrm{LS}}(I - M, \Omega, 0) = \deg_{\mathrm{LS}}(I - H(\cdot, 1), \Omega, 0) \\
&= \deg_{\mathrm{LS}}(I - H(\cdot, 0), \Omega, 0) = \deg_{\mathrm{LS}}(I - P - JQN, \Omega, 0) \\
&= \deg_{\mathrm{B}}((I - P - JQN)\big|_{\mathrm{Ker\,L} \cap \bar{\Omega}}, \mathrm{Ker}\, L \cap \Omega, 0) \\
&= \deg_{\mathrm{B}}(-JQN\big|_{\mathrm{Ker}\, L \cap \bar{\Omega}}, \mathrm{Ker}\, L \cap \Omega, 0) \\
&= (-1)^n \deg_{\mathrm{B}}(JQN\big|_{\mathrm{Ker} L \cap \bar{\Omega}}, \mathrm{Ker} L \cap \Omega, 0) \neq 0,
\end{aligned}$$

where $n = \dim \mathrm{Ker}\, L$, and $\deg_{\mathrm{B}}$ is the Brouwer degree. This is the end.

2.8 Topological Degree of Condensing Fields

2.8.1 Measure of Noncompactness

Definition 2.8.1. Let X be a Banach space and S a bounded subset of X. Denote

$$\alpha(S) = \inf\{\delta > 0;\ S \text{ admits a finite cover by sets of diameter} \leq \delta\}.$$

Then $\alpha(S)$ is called the **Kuratowski measure of noncompactness** of S, in short, **measure of noncompactness** of S.

Clearly, $0 \leq \alpha(S) < \infty$.

Theorem 2.41. *Let S, T be two bounded subsets of X, a be a real number, and $\alpha(S)$ be the measure of noncompactness. Then the following conclusions hold:*

(i) $\alpha(S) = 0 \Longleftrightarrow S$ *is relative compact;*
(ii) $S \subset T \Longrightarrow \alpha(S) \leq \alpha(T)$;
(iii) $\alpha(\bar{S}) = \alpha(S)$;
(iv) $\alpha(S \bigcup T) = \max\{\alpha(S), \alpha(T)\}$;
(v) $\alpha(aS) = |a|\alpha(S)$, *where* $aS = \{x = az;\ z \in S\}$;
(vi) $\alpha(S + T) \leq \alpha(S) + \alpha(T)$, *where* $S + T = \{x = y + z;\ y \in S, z \in T\}$;
(vii) $\alpha(\overline{co}S) = \alpha(S)$, *where* $\overline{co}S$ *is the closed convex hull of* S;
(viii) $|\alpha(S) - \alpha(T)| \leq 2d_h(S, T)$, *where* $d_h(S, T)$ *is the* **Hausdorff distance** *between* S *and* T, *that is,* $d_h(S, T) = \max\{\sup_{x \in S} d(x, T), \sup_{x \in T} d(x, S)\}$, *and* $d(\cdot, \cdot)$ *is the distance between a point and a set.*

Proof. (i) and (ii) are clear.

(iii) Since $S \subset \bar{S}$, then $\alpha(S) \leq \alpha(\bar{S})$.

On the other hand, by the definition, for $\forall \epsilon > 0$, there exists a decomposition $S = \bigcup_{i=1}^m S_i$, such that $\mathrm{diam}(S_i) < \alpha(S) + \epsilon$ $(1 \leq i \leq m)$, where $\mathrm{diam}(S_i)$ is the diameter of S_i. Since $\mathrm{diam}(\bar{S}_i) = \mathrm{diam}(S_i) < \alpha(S) + \epsilon$ and $\bar{S} = \bigcup_{i=1}^m \bar{S}_i$, it follows that $\alpha(\bar{S}) \leq \alpha(S) + \epsilon$. By the arbitrariness of ϵ, we arrive at $\alpha(\bar{S}) \leq \alpha(S)$.

(iv) Let $\eta = \max\{\alpha(S), \alpha(T)\}$. By (ii), $\eta \leq \alpha(S \bigcup T)$.

On the other hand, by definition, $\forall \epsilon > 0$, there exist $S = \bigcup_{i=1}^m S_i$ and $T = \bigcup_{j=1}^n T_j$, such that $\mathrm{diam}(S_i) < \alpha(S) + \epsilon \leq \eta + \epsilon$ and

$\text{diam}(T_j) < \alpha(T) + \epsilon \leq \eta + \epsilon$, $1 \leq i \leq m$, $1 \leq j \leq n$. Since $S \bigcup T = (\bigcup_{i=1}^m S_i) \bigcup (\bigcup_{j=1}^n T_j)$, it follows that $\alpha(S \bigcup T) \leq \eta + \epsilon$. By the arbitrariness of ϵ, we arrive at $\alpha(S \bigcup T) \leq \eta$.

(v) If $a = 0$, it is clear that $\alpha(aS) = |a|\alpha(S)$ holds.

In the sequel, we assume that $a \neq 0$. $\forall \epsilon > 0$, there exists $S = \bigcup_{i=1}^m S_i$ such that $\text{diam}(S_i) < \alpha(S) + \epsilon$ $(1 \leq i \leq m)$. Clearly, $aS = \bigcup(aS_i)$, $\text{diam}(aS_i) = |a|\text{diam}(S_i) < |a|\alpha(S) + |a|\epsilon$. Then $\alpha(aS) \leq |a|\alpha(S) + |a|\epsilon$. By the arbitrariness of ϵ, we arrive at $\alpha(aS) \leq |a|\alpha(S)$.

On the other hand, by the above conclusion, we also have that $\alpha(S) = \alpha(a^{-1}aS) \leq |a^{-1}|\alpha(aS)$. Then $\alpha(aS) \geq |a|\alpha(S)$.

(vi) $\forall \epsilon > 0$, there exist $S = \bigcup_{i=1}^m S_i$ and $T = \bigcup_{j=1}^n T_j$, such that $\text{diam}(S_i) < \alpha(S) + \epsilon$, $\text{diam}(T_j) < \alpha(T) + \epsilon$. Let

$$V_{ij} = \{x; x = y + z, y \in S_i,\ z \in T_j\}.$$

Clearly, $S + T = \bigcup_{ij} V_{ij}$, $\text{diam}(V_{ij}) \leq \text{diam}(S_i) + \text{diam}(T_j) < \alpha(S) + \alpha(T) + 2\epsilon$. Then $\alpha(S + T) \leq \alpha(S) + \alpha(T) + 2\epsilon$. By the arbitrariness of ϵ, we arrive at $\alpha(S + T) \leq \alpha(S) + \alpha(T)$.

(vii) Since $S \subset \overline{co}S$, we only need to prove that $\alpha(\overline{co}S) \leq \alpha(S)$. For $\mu > \alpha(S)$, there exists $S = \bigcup_{i=1}^m S_i$ such that $\text{diam}(S_i) \leq \mu$.

Since $\text{diam}(\overline{co}S_i) \leq \mu$, we can assume that each S_i is a convex subset, without loss of generality. Since

$$\begin{aligned} \overline{co}S &\subset \overline{co}\left[S_1 \bigcup \overline{co}\left(\bigcup_{i=2}^m S_i\right)\right] \\ &\subset \overline{co}\left[S_1 \bigcup \overline{co}\left[S_2 \bigcup \overline{co}\left(\bigcup_{i=3}^m S_i\right)\right]\right] \subset \cdots, \end{aligned}$$

we only need to prove that $\alpha(\overline{co}(C_1 \bigcup C_2)) \leq \max\{\alpha(C_1), \alpha(C_2)\}$, for any bounded convex subsets C_1, C_2.

It is easy to see that $\overline{co}(C_1 \bigcup C_2) \subset \bigcup_{0 \leq \lambda \leq 1}[\lambda C_1 + (1 - \lambda)C_2]$, $C_1 \pm C_2$ are bounded, and there exists $r > 0$ such that $|x| \leq r$, $\forall x \in C_1 \pm C_2$. For any $\epsilon > 0$, we can find $\lambda_1, \cdots, \lambda_p$ such that $[0, 1] \subset \bigcup_{i=1}^p B_{\epsilon/r}(\lambda_i)$. Thus,

$$\overline{co}(C_1 \bigcup C_2) \subset \bigcup_{i=1}^p [\lambda_i C_1 + (1 - \lambda_i)C_2 + \bar{B}_\epsilon(0)].$$

Clearly, $\alpha(\bar{B}_\epsilon(0)) \leq 2\epsilon$. By (v) and (vi), we have

$$\alpha\left(\overline{co}\left(C_1 \bigcup C_2\right)\right) \leq \max\{\alpha(C_1), \alpha(C_2)\} + 2\epsilon, \quad \forall \epsilon > 0.$$

(viii) For any $\epsilon > 0$, there exists $S = \bigcup_{j=1}^m S_j$ such that $\operatorname{diam}(S_j) < \alpha(S) + \epsilon$, $1 \leq j \leq m$. Let

$$N_i = \{y \in T;\ \exists x \in S_i,\ ||x - y|| < 2d_h(S,T)\}.$$

Then

$$\operatorname{diam}(N_i) \leq 2d_h(S,T) + \operatorname{diam}(S_i) \leq 2d_h(S,T) + \alpha(S) + \epsilon.$$

At this moment, it is easy to know that $T \subset \bigcup_{j=1}^m N_j$. So, $\alpha(T) \leq 2d_h(S,T) + \alpha(S) + \epsilon$. By the arbitrariness of ϵ, we arrive at

$$\alpha(T) \leq \alpha(S) + 2d_h(S,T).$$

By the same reason, it can be proven that

$$\alpha(S) \leq \alpha(T) + 2d_h(S,T).$$

Therefore, $|\alpha(S) - \alpha(T)| \leq 2d_h(S,T)$. This is the end.

Lemma 2.15. *Let $\{S_n\}$ be a sequence of bounded nonempty closed subsets in X, $S_1 \supset S_2 \supset \cdots \supset S_n \supset \cdots$, and $\alpha(S_n) \to 0$ as $n \to \infty$. Then $S = \bigcap_{n=1}^\infty S_n$ is a nonempty compact subset in X.*

Proof. It only need to prove that for any sequence $x_n \in S_n\,(n = 1, 2, \ldots)$, there exists a subsequence $\{x_{n_i}\} \subset \{x_n\}$, such that $x_{n_i} \to x_0 \in S$.

Let $\alpha(S_n) =: \alpha_n$. Then S_n has a decomposition $S_n = \bigcup_{i=1}^{p_n} S_i^{(n)}$, where $\operatorname{diam}(S_i^{(n)}) < \alpha_n + \frac{1}{n}$ $(i = 1, 2, \ldots, p_n)$. Since $\{x_n\} \subset S_1$, there must be a subsequence $\{x_n^{(1)}\}$ of $\{x_n\}$ completely contained within a certain $S_i^{(1)}$. Thus, $\operatorname{diam}\{x_n^{(1)}\} < \alpha_1 + 1$. Since $\{x_2^{(1)}, x_3^{(1)}, \ldots\} \subset S_2$, there must be a subsequence $\{x_n^{(2)}\}$ of $\{x_n^{(1)}\}\,(n > 1)$ completely contained within a certain $S_i^{(2)}$. Thus, $\operatorname{diam}\{x_n^{(2)}\} < \alpha_2 + \frac{1}{2}$. By the same reason, since $\{x_3^{(2)}, x_4^{(2)}, \ldots\} \subset S_3$, there must be a subsequence $\{x_n^{(3)}\}$ of $\{x_n^{(2)}\}\,(n > 1)$ completely contained within a certain $S_i^{(3)}$.

Thus, $\operatorname{diam}\{x_n^{(3)}\} < \alpha_3 + \frac{1}{3}$. Continuing this process, the resulting diagonal sequence $\{x_n^{(n)}\}$ will converge. In fact, if $m > n$, we have

$$||x_m^{(m)} - x_n^{(n)}|| < \alpha_n + \frac{1}{n}.$$

Since $\alpha_n \to 0\,(n \to \infty)$, it follows that $\{x_n^{(n)}\}$ is a fundamental sequence in X. So, $x_n^{(n)} \to x_0 \in X$. Since $x_m^{(m)} \in S_n$ as $m \geq n$ and S_n is a closed subset, it follows that $x_0 \in S_n$ $(n = 1, 2, \ldots)$. Hence, $x_0 \in S$. This is the end.

Let X, W be two Banach spaces, $D \subset W$ be a compact subset, and $Y = C(D, X)$ be the space composed of all continuous functions $u : D \to X$, with norm $||u||_0 = \max\{||u(\xi)||;\ \xi \in D\}$.

$B \subset Y$ is called **equi-continuous** on D, if $\forall \epsilon > 0$, there exists $\delta = \delta(\epsilon)$, such that, whenever $||\xi - \eta|| \leq \delta$, $\sup\{||u(\xi) - u(\eta)||;\ u \in B\} < \epsilon$.

Theorem 2.42. *Let X, W be two Banach spaces, $D \subset W$ be a compact subset, and $B \subset C(D, X)$. Then the following conclusions hold:*

(a) $\alpha(B) = \sup_D \alpha(B(\xi))$, *if B is bounded and equi-continuous, where $B(\xi) = \{u(\xi);\ u \in B\}$.*
(b) *B is relative compact if and only if B is equi-continuous, and $B(\xi)$ is relative compact for each $\xi \in D$.*

Proof. (a) For $\mu > \alpha(B)$, we have that $B \subset \bigcup_{i=1}^{p} M_i$ and $\operatorname{diam}(M_i) \leq \mu$. It is easy to see that

$$B(\xi) \subset \bigcup_{i=1}^{p} M_i(\xi), \quad \operatorname{diam}(M_i(\xi)) \leq \mu.$$

Thus, $\alpha(B(\xi)) \leq \mu$. So, $\sup_D \alpha(B(\xi)) \leq \alpha(B)$.

Conversely, since D is compact and B is equi-continuous, for any $\epsilon > 0$, there exist a finite number of points $\xi^1, \ldots, \xi^p \in D$, such that $B(\xi) \subset \bigcup_{i=1}^{p}(B(\xi^i) + B_\epsilon(0))$, $\forall \xi \in D$. Furthermore, for $\mu > \sup_D \alpha(B(\xi))$, there exist $M_1, \ldots, M_m$ such that $\operatorname{diam}(M_j) \leq \mu$ and $\bigcup_{i=1}^{p} B(\xi^i) \subset \bigcup_{j=1}^{m} M_j$. Then, B is in the union of a finite number of sets

$$\{u \in B; \quad u(\xi^1) \in M_{j_1}, \ldots, u(\xi^p) \in M_{j_p}\},$$

where each is with a diameter $\leq \mu+2\epsilon$. So, $\alpha(B) \leq \mu+2\epsilon$. Therefore, $\alpha(B) \leq \sup_D \alpha(B(\xi))$.

(b) "$\Longrightarrow$". Assume that B is relative compact. Then, B is bounded, and for any $\epsilon > 0$, there exist $u_1, \ldots, u_p$ such that $B \subset \bigcup_{i=1}^p B_\epsilon(u_i)$. It is easy to know that $\{u_1, \ldots, u_p\}$ is equi-continuous. Thus, there exists $\delta = \delta(\epsilon)$, such that $\sup\{||u(\xi)-u(\eta)||;\ u \in B\} \leq 3\epsilon$ whenever $||\xi - \eta|| \leq \delta$. So, B is equi-continuous. Finally, since $\alpha(B) = \sup_D \alpha(B(\xi))$, it follows that $\alpha(B(\xi)) = 0$. So, $B(\xi)$ is relative compact.

"$\Longleftarrow$". Since B is equi-continuous, for $\forall\epsilon > 0$, there exists $\delta > 0$, such that $||u(\xi) - u(\eta)|| \leq \epsilon$, $\forall u \in B$, whenever $||\xi - \eta|| \leq \delta$. Since D is compact, there exist a finite number of points $\xi_1, \ldots, \xi_p$, such that, $\forall \eta \in D$, there is a certain i, such that $||\eta - \xi_i|| \leq \delta$. Thus, $||u(\eta) - u(\xi_i)|| \leq \epsilon$. It is known that $B(\xi_i)$ $(i = 1, \ldots, p)$ are bounded, because they are relative compact. So, B is bounded. By (a), we obtain that $\alpha(B) = \sup_D \alpha(B(\xi)) = 0$, which implies that B is relative compact. This is the end.

Theorem 2.43. *Suppose that $H \subset C([a,b], X)$ is bounded and equi-continuous. Then $\alpha(H(t))$ is continuous on $[a,b]$, and*

$$\alpha\left(\left\{\int_a^b x(t)dt;\ x \in H\right\}\right) \leq \int_a^b \alpha(H(t))dt.$$

Proof. Since H is equi-continuous, for any $\epsilon > 0$, there exists $\delta > 0$, such that, for each $u \in H$, $||u(t_1) - u(t_2)|| \leq \epsilon$, whenever $|t_1 - t_2| \leq \delta$. Then,

$$||u(t_2)|| - \epsilon \leq ||u(t_1)|| \leq ||u(t_2)|| + \epsilon, \quad \forall u \in H.$$

For $\mu > \alpha(H(t_2))$, we have that $H(t_2) \subset \bigcup_{i=1}^p M_i$ and $\operatorname{diam}(M_i) \leq \mu$. Thus, we obtain that $H(t_1) \subset \bigcup_{i=1}^p (M_i + B_\epsilon(0))$. So, $\alpha(H(t_1)) \leq \mu + 2\epsilon$. Hence,

$$\alpha(H(t_1)) \leq \alpha(H(t_2)) + 2\epsilon.$$

By the same reason, it follows that $\alpha(H(t_2)) \leq \alpha(H(t_1)) + 2\epsilon$. Therefore, $\alpha(H(t))$ is continuous.

For any n, take $t_i = a + i\frac{b-a}{n}$. For $\mu > \int_a^b \alpha(H(t))dt$ and $x \in H$, there exists N such that, when $n \geq N$, we have

$$\mu > \sum_{i=1}^{n} \alpha(H(t_i))\frac{b-a}{n}, \quad \left\| \int_a^b x(t)dt - \sum_{i=1}^{n} x(t_i)\frac{b-a}{n} \right\| < \epsilon.$$

Take a fixed $n > N$, and for each i, take a fixed $\eta_i > \alpha(H(t_i))$ such that $\mu > \sum_{i=1}^{n} \eta_i \frac{b-a}{n}$. Then,

$$H(t_i) \subset \bigcup_{j=1}^{p_i} M_j^{(i)}, \quad \mathrm{diam}(M_j^{(i)}) \leq \eta_i.$$

From this, it can be inferred that for any $x \in H$, $\sum_{i=1}^{n} x(t_i)\frac{b-a}{n}$ is located within

$$\frac{b-a}{n} \bigcup_{j=1}^{p_1} M_j^{(1)} + \cdots + \frac{b-a}{n} \bigcup_{j=1}^{p_n} M_j^{(n)}.$$

So, for any $x \in H$, $\int_a^b x(t)dt$ is located within

$$\frac{b-a}{n} \bigcup_{j=1}^{p_1} M_j^{(1)} + \cdots + \frac{b-a}{n} \bigcup_{j=1}^{p_n} M_j^{(n)} + B_\epsilon(0).$$

Therefore, we arrive at

$$\alpha\left(\left\{\int_a^b x(t)dt;\ x \in H\right\}\right) \leq \sum_{i=1}^{n} \eta_i \frac{b-a}{n} + 2\epsilon < \mu + 2\epsilon.$$

By the arbitrariness of ϵ, the conclusion holds. This is the end.

2.8.2 Topological Degree of Condensing Fields

This section is a brief introduction, please refer to [10, 13] for details.

Definition 2.8.2. Let $D \subset X$ and $F : D \to X$ be a continuous operator. F will be called k**-set contraction operator** on D, if there exists a constant $k \geq 0$, such that for any bounded subset $S \subset D$, we have

$$\alpha(F(S)) \leq k\alpha(S).$$

If $k < 1$, F will be called the **strict set contraction operator**. F is said to be the **condensing operator**, if $\alpha(F(S)) < \alpha(S)$ whenever $S \subset D$ is bounded and $\alpha(S) > 0$, in other words, $\alpha(F(S)) \geq \alpha(S)$ implies $\alpha(S) = 0$.

Clearly, both completely continuous operators and contraction operators are strict set contraction operators, and strict set contraction operator is condensing operator.

Definition 2.8.3. Let U be the retract of the Banach space X, $V \subset U$ be a bounded relative open subset of U, and $F : \bar{V} \to U$ be completely continuous and have no fixed point on ∂V. Let $r : X \to U$ be a retraction. Take R sufficiently large such that $\bar{V} \subset B_R = \{x \in X; ||x|| < R\}$. The **fixed point index** of F on V with respect to U is defined by

$$i(F, V, U) = \deg(id - F \circ r, B_R \cap r^{-1}(V), 0), \tag{2.28}$$

where the right-hand side is Leray–Schauder degree.

It is easy to know that $F \circ r : r^{-1}(\bar{V}) \to X$ is completely continuous, and $F \circ r$ has no fixed point on $\partial(B_R \cap r^{-1}(V))$. Thus, the Leray–Schauder degree at the right-hand side of (2.28) is well defined. It can be proved that $i(F, V, U)$ in the definition is independent of the selection of R and the retract r.

Definition 2.8.4. Let $\Omega \subset X$ be a bounded open subset, $F : \bar{\Omega} \to X$ be a strict set contraction operator, $f = id - F$, and $0 \notin f(\partial\Omega)$. Denote

$$D_1 = \overline{co}F(\bar{\Omega}),\ D_n = \overline{co}F(D_{n-1} \cap \bar{\Omega}), \quad n = 2, 3, \ldots.$$

If $D_{n_0} \cap \bar{\Omega} = \emptyset$ for some $n = n_0$, one defines the topological degree

$$\deg(f, \Omega, 0) = 0.$$

Suppose that $D_n \cap \bar{\Omega} \neq \emptyset$, $n = 1, 2, \ldots$. Let $D = \bigcap_{n=1}^{\infty} D_n$. By Lemma 2.15, D is nonempty compact convex subset in X. Obviously, $F(D \cap \bar{\Omega}) \subset D$. If F is considered as the operator from $D \cap \bar{\Omega}$ into D, then this operator is completely continuous and has no fixed point on $D \cap \partial\Omega$. At this point, one defines the topological degree

$$\deg(f, \Omega, 0) = i(F, D \cap \Omega, D),$$

If $p \in X \backslash f(\partial\Omega)$, one defines the topological degree

$$\deg(f, \Omega, p) = \deg(f - p, \Omega, 0).$$

Definition 2.8.5. Let $\Omega \subset X$ be a bounded open subset, $F : \bar{\Omega} \to X$ be a condensing operator, $f = id - F$, and $p \in X \backslash f(\partial\Omega)$. It can be shown that $\rho = \inf_{x \in \partial\Omega} ||f(x) - p|| > 0$. Take a strict set contraction operator $G : \bar{\Omega} \to X$, such that $||Fx - Gx|| < \rho$, $\forall x \in \bar{\Omega}$. Let $g = id - G$. Then, it is easy to know that $p \notin g(\partial\Omega)$. Thus, we define the topological degree:

$$\deg(f, \Omega, p) = \deg(g, \Omega, p).$$

Firstly, the strict set compression operator in the definition is easy to obtain, for example, taking $G = kF$ to make $k < 1$ and $1 - k$ sufficiently small. Secondly, it is necessary to prove that the $\deg(f, \Omega, p)$ defined by the above method is independent of the selection of G. For the topological degrees of the strict set contraction field and condensing field, we can also verify that the degrees enjoy the normality, additivity, homotopy invariance, Kronecker existence, excision, boundary value property, connected component property, lack of direction, etc. We can also extend the L-compactness in the definition of coincidence degree to L-condensing. Readers can refer to other works [11]. To illustrate it, we discuss a generalized form of Leray–Schauder fixed point.

Theorem 2.44 (Sadovskii's Fixed Point). *Let D be a bounded convex closed subset in X, and $T : D \to D$ be a condensing mapping. Then T admits at least one fixed point in D.*

Proof. For any fixed point $x_0 \in D$, let

$$Z = \{S;\ x_0 \in S \subset D, S \text{ is a convex closed subset}, T(S) \subset S\}.$$

Clearly, $D \in Z$. Denote $S_0 = \bigcap_{S \in Z} S$. It is easy to know that $x_0 \in S_0 \subset D$, S_0 is also a convex closed set, and $T(S_0) \subset S_0$. Since $\overline{co}\{T(S_0), x_0\} \subset S_0$, it follows that

$$T(\overline{co}\{T(S_0), x_0\}) \subset T(S_0) \subset \overline{co}\{T(S_0), x_0\}.$$

Then, $\overline{co}\{T(S_0), x_0\} \in Z$. Thus, $\overline{co}\{T(S_0), x_0\} = S_0$. So,

$$\alpha(S_0) = \alpha(\overline{co}\{T(S_0), x_0\}) = \alpha(\{T(S_0), x_0\}) = \alpha(T(S_0)).$$

Since T is a condensing mapping, we obtain that $\alpha(S_0) = 0$, which implies that S_0 is a convex compact subset. By the Leray–Schauder fixed point, T admits at least one fixed point in S_0. This is the end.

2.9 The Global Bifurcation Theorem

Let X be a real Banach space, $\Lambda = \mathbb{R}$ a parameter space, and $U \subset X \times \Lambda$ an open subset. In this section, we consider the following equation

$$f(x, \lambda) = x - \lambda Ax + g(x, \lambda) = 0, \tag{2.29}$$

where $A : X \to X$ is a completely continuous linear operator, $g : U \to X$ is a completely continuous nonlinear operator, and the limit

$$\lim_{||x|| \to 0} \frac{g(x, \lambda)}{||x||} = 0$$

holds uniformly for λ on any bounded subset in $\mathbb{R}$.

Clearly, $f(0, \lambda) = 0$, $\forall \lambda$. According to the implicit function theorem, if $(0, \lambda_0)$ is a bifurcation point of (2.29), then λ_0^{-1} is an eigenvalue of A.

Theorem 2.45 (Krasnoselskii). *Under the above conditions, if λ_0^{-1} is the odd algebraic multiplicity eigenvalue of A, then $(0, \lambda_0)$ must be a bifurcation point of the equation (2.29).*

Proof. Since A is the completely continuous operator, then its nonzero eigenvalues are isolated. Thus, there exists $\epsilon_0 > 0$, such that A has only a unique eigenvalue ${\lambda_0}^{-1}$ in $[\frac{1}{\lambda_0 + \epsilon_0}, \frac{1}{\lambda_0 - \epsilon_0}]$.

We use the method of contradiction. If $(0, \lambda_0)$ is not bifurcation point of (2.29), then there exist ρ and ϵ_1 such that, whenever $|\lambda - \lambda_0| \leq \epsilon_1$, the equation (2.29) has only the zero solution in $B_X(\rho)$. If $0 < \epsilon_2 < \min\{\epsilon_0, \epsilon_1\}$, by the compact homotopy invariance of L-S degree, we have

$$\deg(f(\cdot, \lambda_0 - \epsilon_2), B_X(\rho), 0) = \deg(f(\cdot, \lambda_0 + \epsilon_2), B_X(\rho), 0). \quad (2.30)$$

From Theorem 2.24, it follows that

$$\begin{aligned}\deg(f(\cdot, \lambda_0 - \epsilon_2), B_X(\rho), 0) &= \operatorname{index}(f(\cdot, \lambda_0 - \epsilon_2), 0)\\ &= \operatorname{index}(id - (\lambda_0 - \epsilon_2)A, 0) = (-1)^{\beta},\end{aligned}$$

where β is the sum of algebraic multiplicity of all eigenvalues of A within $(\frac{1}{\lambda_0 - \epsilon_2}, \infty)$. Similarly, we also have

$$\begin{aligned}\deg(f(\cdot, \lambda_0 + \epsilon_2), B_X(\rho), 0) &= \operatorname{index}(f(\cdot, \lambda_0 + \epsilon_2), 0)\\ &= \operatorname{index}(id - (\lambda_0 + \epsilon_2)A, 0) = (-1)^{\beta + \beta_1},\end{aligned}$$

where β_1 is the algebraic multiplicity of the eigenvalue λ_0^{-1} of A. By (2.30), we arrive at

$$(-1)^{\beta + \beta_1} = (-1)^{\beta}.$$

This contradicts the fact that β_1 is odd. This is the end.

The oddness of the algebraic multiplicity of λ_0^{-1} is a sufficient condition for bifurcation point, but not necessary (see [3]).

In the sequel, we will introduce the global bifurcation theorem due to Rabinowitz, which will be proved using the following lemma [3,28].

Lemma 2.16. *Let M be a compact metric space, A, B be disjoint closed subsets of M. Then the following alternatives hold: either*

(1) *there exists a connected component of M intersecting A and B, or*
(2) *there exist disjoint compact subsets M_1, M_2, such that $M = M_1 \cup M_2$ and*

$$A \subset M_1, \ B \subset M_2.$$

Theorem 2.46 (P. H. Rabinowitz). *Under the above conditions, let λ_0^{-1} be the odd algebraic multiplicity eigenvalue of A, and $U \subset X \times \mathbb{R}$ be the open subset containing $(0, \lambda_0)$. Denote*

$$S = \{(x, \lambda) \in U;\ f(x, \lambda) = 0,\ x \neq 0\}.$$

Then, the connected component C of $\bar{S}$ containing $(0, \lambda_0)$ possesses one of the following properties:

(1) $C \cap \partial U \neq \emptyset$.
(2) *C is unbounded in U.*
(3) *C contains odd number of trivial points $(0, \lambda_i) \neq (0, \lambda_0)$, where λ_i^{-1} is the odd algebraic multiplicity eigenvalue of A.*

Proof. Suppose that C does not possess the properties (1) and (2). Then C is a bounded closed set. Since A and g are compact, then C is a compact subset in U. Since A is a compact linear operator, C can contain at most a finite number of points $(0, \lambda)$, where λ^{-1} is the eigenvalue of A, denoted as $(0, \lambda_j)$, $j = 0, 1, 2, \ldots, k$. Let

$$N_\delta(C) = \{(x, \lambda) \in U;\ \mathrm{dist}((x, \lambda), C) < \delta\}.$$

Take δ sufficiently small, such that when λ^{-1} is the eigenvalue of A and $(0, \lambda) \neq (0, \lambda_j)$, $j = 0, 1, \ldots, k$, we have $(0, \lambda) \notin \overline{N_\delta(C)}$.

If $(\partial N_\delta(C)) \cap \bar{S} = \emptyset$, then let $\Omega = N_\delta(C)$.

If $(\partial N_\delta(C)) \cap \bar{S} \neq \emptyset$, denote $C_0 = (\partial N_\delta(C)) \cap \bar{S}$. Thus, C_0 is closed and $C_0 \cap C = \emptyset$. By Lemma 2.16, there exist compact sets C_1, C_2 such that $C_1 \supset C, C_2 \supset C_0$, $\overline{N_\delta(C)} \cap \bar{S} = C_1 \cup C_2$, and $\mathrm{dist}(C_1, C_2) = \rho > 0$. At this point, let $\Omega = N_\delta(C) \cap N_{\frac{\rho}{2}}(C_1)$. Then, $\bar{\Omega} \cap \bar{S} = C_1$ and $\partial\Omega \cap \bar{S} = \emptyset$.

In this way, the bounded open subset Ω of U can be builded, such that

(i) $C \subset \Omega$.
(ii) For any $(x, \lambda) \in \partial\Omega$, if $f(x, \lambda) = 0$, then $x = 0$.
(iii) Except for $(0, \lambda_j)$, $j = 0, 1, \ldots, k$, for any other eigenvalue λ^{-1} of A, there is $(0, \lambda) \notin \Omega$.

Consider auxiliary mapping $f_\rho : \bar{\Omega} \to X \times \mathbb{R}$,

$$f_\rho(x, \lambda) = (f(x, \lambda), ||x||^2 - \rho^2),$$

where ρ is parameter.

It is easy to see that f_ρ is a completely continuous field, and the necessary and sufficient condition for $(\bar{x}, \bar{\lambda}) \in \Omega$ to be the zero point of f_ρ is that $(\bar{x}, \bar{\lambda})$ is the zero point of f and $||\bar{x}|| = \rho$.

By (ii), it is known that for any $\rho > 0$, f_ρ has no zero point on $\partial\Omega$. So, f_ρ is compact homotopy, and $\deg(f_\rho, \Omega, 0)$ is independent of ρ.

If ρ is sufficiently large, it follows from the boundedness of Ω that f_ρ has no zero point on Ω, which implies $\deg(f_\rho, \Omega, 0) = 0$.

Let's examine the expression of $\deg(f_\rho, \Omega, 0)$ when ρ is sufficiently small.

Take ϵ sufficiently small such that, whenever $|\lambda - \lambda_j| \leq \epsilon$ and $\lambda \neq \lambda_j\, (j = 0, 1, \ldots, k)$, λ^{-1} is not the eigenvalue of A. Put

$$\Omega_j = \{(x, \lambda) \in \Omega;\ ||x||^2 + (\lambda - \lambda_j)^2 \leq \rho^2 + \epsilon^2\}.$$

Then, when ρ is sufficiently small and does not equal to zero, f_ρ does not have any zero point on $\Omega \backslash (\bigcup_{j=0}^{\infty} \Omega_j)$.

In fact, by the property (iii) of Ω, there exists a constant $r > 0$ such that, whenever $(0, \lambda) \in \Omega$ and $|\lambda - \lambda_j| \geq \epsilon$, $j = 0, 1, \ldots, k$, we have

$$||(id - \lambda A)^{-1}|| \leq r. \tag{2.31}$$

Based on the assumption of g, when $||x|| \leq \rho$ and ρ is sufficiently small, we have

$$||g(x, \lambda)|| \leq \frac{1}{2r}||x||. \tag{2.32}$$

By contradiction, suppose $f_\rho(x, \lambda) = 0$ for some $(x, \lambda) \in \Omega \backslash (\bigcup_{j=0}^{\infty} \Omega_j)$. Then, we have $(id - \lambda A)x + g(x, \lambda) = 0$, $||x|| = \rho$ and $|\lambda - \lambda_j| \geq \epsilon$, $j = 0, 1, \ldots, k$. By combining (2.31) and (2.32), we obtain that

$$\rho = ||x|| = ||(id - \lambda A)^{-1} g(x, \lambda)|| \leq \frac{1}{2}||x||,$$

which is a contradiction. So, f_ρ has no zero point on $\Omega \backslash (\bigcup_{j=0}^{\infty} \Omega_j)$.

By the excision of the degree, we arrive at

$$\deg(f_\rho, \Omega, 0) = \sum_{j=0}^{k} \deg(f_\rho, \Omega_j, 0).$$

In the sequel, we will compute $\deg(f_\rho, \Omega_j, 0)$. For this purpose, we make the homotopy as follows:

$$h_t(x,\lambda) = ((id-\lambda A)x + tg(x,\lambda), t(||x||^2-\rho^2)+(1-t)(\epsilon^2-(\lambda-\lambda_j)^2)).$$

Then, whenever $(x,\lambda) \in \partial\Omega_j$ and $t \in [0,1]$, $h_t(x,\lambda) \neq 0$.

In fact, by contradiction, suppose that there exist some $t \in [0,1]$ and some $(x,\lambda) \in \partial\Omega_j$, such that $h_t(x,\lambda) = 0$. Then,

$$\begin{cases} (id-\lambda A)x + tg(x,\lambda) = 0, \\ t(||x||^2-\rho^2)+(1-t)(\epsilon^2-(\lambda-\lambda_j)^2) = 0. \end{cases} \tag{2.33}$$

Since $(x,\lambda) \in \partial\Omega_j$, it follows that $||x||^2 + |\lambda-\lambda_j|^2 = \rho^2+\epsilon^2$. Thus, from the second equation of (2.33), we can obtain $\lambda = \lambda_j \pm \epsilon$. At this point, by using (2.31) and (2.32), we can obtain

$$(id-\lambda A)x + tg(x,\lambda) \neq 0,$$

which is in contradiction with the first equation of (2.33).

Hence, by the compact homotopy invariance of Leray–Schauder degree, we arrive at

$$\deg(f_\rho, \Omega_j, 0) = \deg(h_1, \Omega_j, 0) = \deg(h_0, \Omega_j, 0).$$

It is easily known that $h_0(x,\lambda) = ((id-\lambda A)x, \epsilon^2-(\lambda-\lambda_j)^2)$ has only two isolated regular zeros $(x,\lambda) = (0, \lambda_j \pm \epsilon)$ in Ω_j, and

$$h_0'(0,\lambda_j \pm \epsilon)(h,\mu) = ((id-(\lambda_j \pm \epsilon)A)h, -2(\pm\epsilon)\mu).$$

From the index formula of regular zeros, it can be obtained

$$\begin{aligned} \text{index}(h_0,(0,\lambda_j+\epsilon)) &= -\text{index}(id-(\lambda_j+\epsilon)A, 0) =: -i_j^+; \\ \text{index}(h_0,(0,\lambda_j-\epsilon)) &= \text{index}(id-(\lambda_j-\epsilon)A, 0) =: i_j^-. \end{aligned}$$

So,

$$\deg(h_0, \Omega_j, 0) = i_j^- - i_j^+.$$

Therefore, we have that

$$0 = \deg(f_\rho, \Omega, 0) = \sum_{j=0}^{k} \deg(f_\rho, \Omega_j, 0) = \sum_{j=0}^{k} (i_j^- - i_j^+).$$

It is easy to see that when λ_j^{-1} is the even algebraic multiplicity eigenvalue of A, $i_j^- = i_j^+$; when λ_j^{-1} is the odd algebraic multiplicity eigenvalue of A, $i_j^- = -i_j^+$. So, among $\lambda_0^{-1}, \ldots, \lambda_k^{-1}$, there must be an even number of odd algebraic multiplicity eigenvalues of A, and the number of odd algebraic multiplicity eigenvalues with $i_j^- = 1$, $i_j^- = -1$ each accounts for half. This is the end.

Exercises

1. (a) Let $\Omega \subset \mathbb{R}$ be an open interval, $0 \in \Omega$, $f(x) = \alpha x^k$, $\alpha \neq 0$. Show that if k is even number, then $\deg(f, \Omega, 0) = 0$; if k is odd number, then $\deg(f, \Omega, 0) = \operatorname{sign} \alpha$.
   (b) Let $g(x) = f(x) + \sum_{i=0}^{k-1} \alpha_i x^i$, $x \in \mathbb{R}$, where f is as in (a). Show that $\deg(g, (-r, r), 0) = \deg(f, (-r, r), 0)$, for sufficiently large r.
2. Let $\Omega = \{(x, y);\ x^2 + y^2 < 1\}$ be a subset in $\mathbb{R}^2$. Denote the mappings f, g by
$$f(x, y) = (e^x - 1, y^2),\ g(x, y) = (y - x^3, y).$$
   Compute $\deg(f, \Omega, (0, 0))$ and $\deg(g, \Omega, (0, 0))$.
3. Let $\Omega \subset \mathbb{R}^n$ be a bounded open subset, $f \in C(\bar{\Omega})$, $g \in C(\bar{\Omega})$, and $|g(x)| < |f(x)|$ on $\partial\Omega$. Show that $\deg(f + g, \Omega, 0) = \deg(f, \Omega, 0)$.
4. Show that the system $2x + y + \sin(x + y) = 0$, $x - 2y + \cos(x + y) = 0$ has a solution in $B_r(0) \subset \mathbb{R}^2$, where $r > 1/\sqrt{5}$.
5. Let $\Omega = B_1(0) \subset \mathbb{R}^n$, $f \in C(\bar{\Omega})$, and $0 \notin f(\bar{\Omega})$. Show that there exist $x, y \in \partial\Omega$ and $\lambda > 0$, $\mu < 0$ such that $f(x) = \lambda x$ and $f(y) = \mu y$, that is, f has a positive and negative eigenvalue, each with an eigenvector in $\partial\Omega$.
6. In finite dimensional space, show that the spherical surface ∂B is not retract of the ball B, that is, there is no continuous mapping $f : B \to \partial B$, such that $f(x) = x$, $\forall x \in \partial B$.
7. Let $\Omega = B_1(0) \subset \mathbb{R}^{2m+1}$, $f : \partial\Omega \to \partial\Omega$ be continuous. Show that there exists $x \in \partial\Omega$ such that either $x = f(x)$ or $x = -f(x)$ holds.
8. Let $f \in C(\mathbb{R}^n)$ satisfy $(f(x), x)/|x| \to \infty$(as $|x| \to \infty$). Show that $f(\mathbb{R}^n) = \mathbb{R}^n$.

9. Let A be a real $n \times n$ matrix, $\det A \neq 0$, and $f \in C(\mathbb{R}^n)$, such that $|x - Af(x)| \leq \alpha|x| + \beta$ on $\mathbb{R}^n$, where $\alpha \in [0, 1)$ and $\beta \geq 0$. Show that $f(\mathbb{R}^n) = \mathbb{R}^n$.
10. Let $f \in C(\mathbb{R}^n)$ be such that f maps $\partial B_r(0)$ onto itself and is surjective, where $r > 0$. Show that $\deg(f^m, B_r(0), 0) = [\deg(f, B_r(0), 0)]^m$.
11. Let $\Omega \subset \mathbb{R}^n$ be a bounded open subset, $f \in C(\bar{\Omega})$, $f(\bar{\Omega}) \subset \bar{\Omega}$ and $f(x) = x$ on $\partial\Omega$. Show that $f(\bar{\Omega}) = \bar{\Omega}$.
12. Let $\Omega \subset \mathbb{R}^n$ be a bounded open subset, $0 \in \Omega$, $f \in C(\bar{\Omega})$, and $(f(x), x) \geq 0$ on $\partial\Omega$. Show that f has a zero point.
13. Let Ω_m, Ω_n be bounded open subsets in $\mathbb{R}^m$, $\mathbb{R}^n$, respectively, $f \in C(\bar{\Omega}_m, \mathbb{R}^m)$, $g \in C(\bar{\Omega}_n, \mathbb{R}^n)$, $y \in \mathbb{R}^m \backslash f(\partial\Omega_m)$, $z \in \mathbb{R}^n \backslash f(\partial\Omega_n)$. Show that
$$\deg((f, g), \Omega_m \times \Omega_n, (y, z)) = \deg(f, \Omega_m, y)\deg(g, \Omega_n, z).$$
14. Consider $u' = f(t, u)$, where $f \in C^1(\mathbb{R} \times \mathbb{R}^n)$ and f is ω-periodic in t. Suppose that $\langle \operatorname{grad} \phi(x), f(t, x) \rangle \geq 0$, for all $t \in [0, \omega]$ and $|x| \geq \rho$, where $\phi : \mathbb{R}^n \to \mathbb{R}^n$ is continuously differentiable and $\phi(x) \to -\infty$ as $|x| \to \infty$. Show that $u' = f(t, u)$ has an ω-periodic solution.
15. Let $f : \mathbb{R} \times \mathbb{R}^n \to \mathbb{R}^n$ be continuous, ω-periodic with respect to the first variable, and satisfy $\langle f(t, x) - f(t, y), x - y \rangle \geq c|x - y|^2$, for some $c > 0$. Show that $x'' = f(t, x)$ has a unique ω-periodic solution.
16. (Lusternik–Schnirelmann–Borsuk) Let $\Omega \subset \mathbb{R}^n$ be bounded open set and symmetric with respect to the origin, and let $\{A_1, A_2, \ldots, A_p\}$ be a covering of $\partial\Omega$, where $A_i \subset \partial\Omega$ is closed subset such that $A_i \cap (-A_i) = \emptyset$, $i = 1, 2, \ldots, p$. Show that $p \geq n + 1$.
17. Let $J = [0, \pi]$, $X = C(J)$. $F : X \to X$ is defined by
$$(Fx)(t) = \frac{2}{\pi} \int_0^\pi (a \sin t \sin s + b \sin 2t \sin 2s)[x(s) + x^3(s)]ds.$$
Compute $F'(x)$, the eigenvalue of $F'(0)$ and its algebraic multiplicity.

18. Let Ω be a bounded open subset of Banach space X containing the origin, $F : \bar{\Omega} \to X$ be completely continuous and satisfy

$$||x - F(x)||^2 \geq ||F(x)||^2 - ||x||^2, \quad \forall x \in \partial\Omega.$$

Show that F must have a fixed point on $\bar{\Omega}$.

19. Let Ω be a bounded open subset of Banach space X, K be a closed cone in X, $0 \in \Omega$, $F : K \cap \bar{\Omega} \to K$ be completely continuous, $F(0) = 0$, and satisfy

$$\inf_{x \in K \cap \partial\Omega} ||F(x)|| > 0.$$

Show that there exist $x_0 \in K \cap \partial\Omega$ and $\mu_0 > 0$, such that $F(x_0) = \mu_0 x_0$. At this point, μ_0 is also called the eigenvalue and x_0 is the eigenvector.

20. Let X be a Banach space, $f = id - F : X \to X$ be a completely continuous field and satisfy $||f(x) - f(y)|| \geq \phi(||x - y||)$, where $\phi : (0, \infty) \to (0, \infty)$ is continuous and satisfies that if $\phi(r) \to 0$, then $r \to 0$. Show that f is a homeomorphism on X.

21. Consider the integral equation

$$\phi(x) = \int_G k(x, y, \phi(y))dy,$$

where G is a bounded closed subset in $\mathbb{R}^n$. Suppose that $k(x, y, u)$ is continuous on $x \in G, y \in G, -\infty < u < \infty$, and satisfies the following inequality

$$|k(x, y, u)| \leq a + b|u|, \quad \forall x, y \in G, \ -\infty < u < \infty,$$

where $a > 0, b > 0, b \operatorname{mes} G < 1$. Show that the integral equation must have a continuous solution.

22. Let X be an infinite real Banach space, B_1 and S_1 be the unit ball and unit spherical surface, respectively, that is, $B_1 = \{x;\ ||x|| < 1\}$, $S_1 = \{x;\ ||x|| = 1\}$. Show that $\alpha(B_1) = \alpha(S_1) = 2$.

Chapter 3

Variational Methods

Variational calculus was born around the same time as calculus and has a history of more than 300 years. It is an important component of mathematical analysis, closely related to other branches of mathematics, and has a wide range of applications. The initial main content was to reduce the solution of variational problems to differential equations. Since the 1970s, scholars have been seeking theories for other types of critical points beyond the minimizer (maximizer), and critical point theory has made significant progress and achieved important applications in partial differential equations and dynamical systems with variational structures.

3.1 Extremum Principle

3.1.1 Necessary Conditions for Extremum

Theorem 3.1. *Suppose that f is a differentiable functional on Banach space X. Then the necessary condition for f to reach a local minimum at $x_0 \in X$ is that $f'(x_0) = 0$.*

Proof. Since f reaches a local minimum at x_0, *i.e.*, there exists $\delta > 0$ such that $f(x) \geq f(x_0)$ whenever $||x - x_0|| < \delta$, it follows that for any $h \in X$, $||h|| = 1$, and $|t| < \delta$, we have

$$f(x_0 + th) - f(x_0) = f'(x_0)(th) + o(|t|) \geq 0.$$

Because t can be positive or negative, we obtain that $f'(x_0)h = 0$, $\forall h \in X$. This is the end.

3.1.2 Euler–Lagrange Equation

Let's first introduce a lemma, whose proof and some weaker forms can be found in [7].

Lemma 3.1 (du Bois–Reymond). (1)*Let* $\psi \in C([t_0, t_1])$ *and*

$$\int_J \psi(t)\dot{\phi}(t)dt = 0, \quad \forall \phi \in C_0^1(J),$$

where $C_0^1(J) = \{u \in C^1(J);\ u(t_0) = u(t_1) = 0\}$, $J = [t_0, t_1]$. *Then* ψ *is constant.*

(2) *Let* $f \in L^1([t_0, t_1])$ *satisfy*

$$\int_J f(t)\dot{\phi}(t)dt = 0, \quad \forall \phi \in C_0^\infty([t_0, t_1]).$$

Then $f(t) = c$, *a.e.* $t \in [t_0, t_1]$.

First, we discuss the case where the functional depends on one variable functions. Let $J = [t_0, t_1] \subset \mathbb{R}$ be an interval, $L = L(t, u, p)$ be a continuously differentiable function, and $L \in C^1(J \times \mathbb{R}^N \times \mathbb{R}^N, \mathbb{R})$. Take two points $P_0, P_1 \in \mathbb{R}^N$. Denote

$$M = \{u \in C^1(J, \mathbb{R}^N);\ u(t_i) = P_i,\ i = 0, 1\}.$$

Define the functional on M by

$$I(u) = \int_J L(t, u(t), \dot{u}(t))dt.$$

Suppose that u^* is a local minimizer of I on M. In the sequel, we will seek the necessary conditions that must be met for the functional I to reach a minimum at the function u^*.

Similar to the proof of the necessary conditions for extremum above, for any $\phi \in C_0^1(J, \mathbb{R}^N)$, there exists $\epsilon = \epsilon(\phi) > 0$ such that, when $0 < |\epsilon| < \epsilon(\phi)$, we have

$$I(u^* + \epsilon\phi) - I(u^*) \geq 0.$$

Then, we obtain that

$$\begin{aligned}
\delta I(u^*, \phi) &= \lim_{\epsilon \to 0^+} \frac{1}{\epsilon}(I(u^* + \epsilon\phi) - I(u^*)) \\
&= \int_J \sum_{k=1}^{N} [L_{u^k}(t, u^*(t), \dot{u}^*(t))\phi^k(t) + L_{p^k}(t, u^*(t), \dot{u}^*(t))\dot{\phi}^k(t)]dt \\
&= -\int_J \sum_{k=1}^{N} \left(\int_{t_0}^{t} L_{u^k}(s, u^*(s), \dot{u}^*(s))ds \right. \\
&\quad \left. - L_{p^k}\left(t, u^*(t), \dot{u}^*(t)\right) \right) \dot{\phi}^k(t)dt \\
&\geq 0,
\end{aligned}$$

$\forall \phi \in C_0^1([t_0, t_1])$. $\delta I(u^*, \phi)$ is said to be the **first variation** of I with respect to ϕ. By replacing ϕ with $-\phi$, we obtain

$$\int_J \sum_{k=1}^{N} \left(\int_{t_0}^{t} L_{u^k}(s, u^*(s), \dot{u}^*(s))ds - L_{p^k}(t, u^*(t), \dot{u}^*(t)) \right) \dot{\phi}^k(t)dt = 0.$$

According to the du Bois–Reymond Theorem, $u^*(t)$ satisfies the following equation (referred to as the **Euler–Lagrange equation in integral form**, for short, **E–L equation**):

$$\begin{aligned}
&\int_{t_0}^{t} L_{u^k}(s, u^*(s), \dot{u}^*(s))ds - L_{p^k}(t, u^*(t), \dot{u}^*(t)) = \text{const.}, \\
&\quad 1 \leq k \leq N, \ \forall t.
\end{aligned}$$

The integral form of the E–L equation can be rewritten as a differential form of the E–L equation:

$$-DL_p(t, u(t), \dot{u}(t)) + L_u(t, u(t), \dot{u}(t)) = 0,$$

where D denotes weak derivative. In particular, if $L \in C^2$ and $u \in C^2$, then D can be replaced by the ordinary derivative $\frac{d}{dt}$. The E–L equation is a necessary condition for functional minimum, but it is not sufficient.

To discuss sufficient conditions, we consider the **second variation** of I with respect to ϕ at u^* (assuming $L \in C^2(J \times \mathbb{R}^N \times \mathbb{R}^N)$)

$$\begin{aligned}\delta^2 I(u^*, \phi) &= \frac{d^2}{ds^2} I(u^* + s\phi)\Big|_{s=0} \\ &= \sum_{i,j} \int_J [L_{u^i u^j}(t, u^*(t), \dot{u}^*(t))\phi^i(t)\phi^j(t) \\ &\quad + 2L_{u^i p^j}(t, u^*(t), \dot{u}^*(t))\phi^i(t)\dot{\phi}^j(t) \\ &\quad + L_{p^i p^j}(t, u^*(t), \dot{u}^*(t))\dot{\phi}^i(t)\dot{\phi}^j(t)]dt.\end{aligned}$$

The following theorem has been shown in [7].

Theorem 3.2. *Let $L \in C^2(J \times \Omega \times \mathbb{R}^N)$. Suppose that $u^* \in M$ satisfies the E-L equation, and there exists a $\lambda > 0$ such that*

$$\delta^2 I(u^*, \phi) \geq \lambda \int_J |\dot{\phi}|^2 dt, \quad \forall \phi \in C_0^1(J, \mathbb{R}^N).$$

Then u^* is a strict minimizer of I.

The function we discussed earlier is one variable function, and similar conclusions can be drawn when the function is multi-variables function. Let $\Omega \subset \mathbb{R}^n$ be a bounded domain, where its boundary $\partial\Omega \in C^1$. Suppose that the Lagrange function $L = L(x, u, p) \in C^2(\bar{\Omega} \times \mathbb{R}^N \times \mathbb{R}^N)$, and there is a function $\Phi \in C^1(\partial\Omega, \mathbb{R}^N)$ defined on the boundary. On $M := \{v \in C^1(\bar{\Omega}, \mathbb{R}^N);\ v|_{\partial\Omega} = \Phi|_{\partial\Omega}\}$, the minimizer $u^*(x)$ of the functional

$$I(u) = \int_\Omega L(x, u(x), \nabla u(x))dx$$

satisfies the following Euler–Lagrange equation (for short, E–L equation):

$$\sum_{\alpha=1}^n \frac{\partial L_{p_\alpha^i}(x, u^*(x), \nabla u^*(x))}{\partial x_\alpha} - L_{u^i}(x, u^*(x), \nabla u^*(x)) = 0, \quad 1 \leq i \leq N.$$

It is easily known that if $N = 1$ and $L(p) = \frac{1}{2}|p|^2$, on $M := \{v \in C^1(\bar{\Omega}, \mathbb{R}^N); v\big|_{\partial\Omega} = \Phi\big|_{\partial\Omega}\}$, the minimizer of the functional

$$D(u) = \frac{1}{2}\int_\Omega |\nabla u(x)|^2 dx \tag{3.1}$$

satisfies the following E–L equation

$$\Delta u = \sum_{\alpha=1}^{n} \frac{\partial^2 u}{\partial x_\alpha^2} = 0, \quad \forall x \in \Omega \tag{3.2}$$

and the boundary condition

$$u\big|_{\partial\Omega} = \Phi. \tag{3.3}$$

The equation (3.2) is called the **harmonic equation**, and the boundary condition (3.3) is called the **Dirichlet boundary condition**.

This idea opens a new approach to solve differential equations. If a differential equation is an E–L equation of a functional, then we can conversely reduce the problem of solving the differential equation to that of finding the extremum of the corresponding functional. For the above example, in order to prove that the harmonic equation and boundary value problem (3.2)+(3.3) has a solution, it is sufficient to prove that the Dirichlet integral functional (3.1) has a minimizer on M.

Before the 20th century, variational theory was based on the E–L equation, and solving functional extremum could be reduced to solving the corresponding E–L equation. The E–L equation is a differential equation. When $n = 1$, it is an ordinary differential equation (or system of ordinary differential equations), and only in some special cases their analytical solutions can be obtained. When $n > 1$, the E–L equation is a partial differential equation, and it is extremely rare to be able to write down the analytical solution of a partial differential equation. Therefore, we need to directly study the existence of extremum of the functionals (known as **direct method**).

3.1.3 The Conditions for the Existence of Extremum

One way to prove the existence of extremum of functionals is to prove that all minimization sequences have subsequences that converge to the minimizer. For this, we need the following concepts.

Definition 3.1.1. $f : X \to \mathbb{R} \cup \{\infty\}$ is said to be **coercive**, if $\lim_{||x||\to\infty} f(x) = \infty$. If $\liminf_{n\to\infty} f(x_n) \geq f(x_0)$ as $x_n \rightharpoonup x_0$ (weak), then $f : X \to \mathbb{R} \cup \{\infty\}$ is said to be **weakly lower semi-continuous**. The set $M \subset X$ is said to be **weakly closed**, if $x_n \in M, x_n \rightharpoonup x_0$ then $x_0 \in M$.

Theorem 3.3. *Let X be a reflexive Banach space, and $M \subset X$ be a weakly closed nonempty subset. If $f : M \to \mathbb{R} \cup \{\infty\}$, $f \not\equiv \infty$ is a weakly lower semi-continuous and coercive function, then f has a minimizer on M.*

Proof. Take a minimizing sequence $\{x_n\} \subset M$ of f, such that

$$\lim_{n\to\infty} f(x_n) = \inf_{x\in M} f(x).$$

Since f is coercive, it follows that $\{x_n\}$ is bounded. Then, $\{x_n\}$ has a weakly convergent subsequence $x_{n_k} \rightharpoonup x_0$. Since M is weakly closed, then $x_0 \in M$. Since f is weakly lower semi-continuous, we have

$$f(x_0) \leq \liminf f(x_{n_k}).$$

Therefore,

$$f(x_0) = \inf_{x\in M} f(x).$$

This is the end.

Remark 1. We recall the Dirichlet integral (3.1). On the one hand, the boundedness of the Dirichlet integral $D(u)$ cannot derive the boundedness in C^1-norm. Even if a sequence is bounded in C^1-norm, there may not necessarily be a convergent subsequence in C^1-norm. On the other hand, we do not know whether $C^1(\bar{\Omega})$ is a dual space of some linear normed space.

The following (semi) norms are closely related to the Dirichlet integral $D(u)$:

$$||u|| = \left(\int_\Omega | \nabla u|^2 dx\right)^{\frac{1}{2}},$$

and

$$||u||_1 = \left[\int_\Omega (\nabla u|^2 + |u|^2) dx\right]^{\frac{1}{2}}.$$

But, the linear space $C^1(\bar{\Omega})$ is not complete by using the above norm. The completeness of this linear space is denoted by $H^1(\Omega) := W^{1,2}(\Omega)$, which is a Hilbert space. According to the Poincaré inequality, $D(u)$ is a norm of $C_0^1(\Omega)$. The closure of $C_0^1(\Omega)$ in $H^1(\Omega)$ is denoted by $H_0^1(\Omega)$, which is a closed subspace of Hilbert space $H^1(\Omega)$.

This situation is universal when solving variational problems by the direct methods, because functionals are variational integrals with derivatives. The C-type norm of the C-class space with the same order of derivatives is determined by the maximum value of the norm of function values and its derivatives point by point, and the C-type norm cannot be controlled by this variational integral. In addition, in order to have weakly sequential compactness, the space must be taken as a dual space of a linear normed space, which should at least be complete. The Sobolev space $W^{1,q}(\Omega)$ is the function space that satisfies these requirements. Since the weak compactness plays an important role in the calculus of variations, that is why we always choose Sobolev space $W^{1,q}$ as working space.

Remark 2. As we have seen, the weakly lower semi-continuous condition plays an important role in the proof of the existence of a minimizer. We consider the level set $f_c = \{x \in X;\ f(x) \leq c\}$, $\forall c \in \mathbb{R}$. We can prove that f is weakly lower semi-continuous $\Longleftrightarrow$ $f_c\,(\forall c \in \mathbb{R})$ is weakly closed. The closed set and the weakly closed set are two different concepts. A weakly closed set must be a closed set, and the converse may not necessarily hold true. But, by the Mazur theorem in a Banach space, we can show that if $C \subset X$ is a convex subset of the Banach space X, then C is closed $\Longleftrightarrow$ C is weakly closed. There are also some theorems about the weakly lower semi-continuity of functional f. We present a theorem, which can be found in other books [7] for more details.

Theorem 3.4 (Tonolli–Morrey). *Suppose that* $L : \bar{\Omega} \times \mathbb{R}^N \times \mathbb{R}^N \to \mathbb{R}$ *satisfies the following conditions:*

(1) $L \in C^1(\bar{\Omega} \times \mathbb{R}^N \times \mathbb{R}^N)$,
(2) $L \geq 0$,
(3) $\forall (x, u) \in \Omega \times \mathbb{R}^N$, $p \to L(x, u, p)$ *is convex.*

Then $I(u) = \int_\Omega L(x, u(x), \nabla u(x))dx$ *is sequentially weakly lower semi-continuous on* $W^{1,q}(\Omega, \mathbb{R}^N)$ $(1 \leq q < \infty)$.

One application of the above theorem is to reduce the existence of solutions to an E–L equation to seeking the extremum of the functional. The minimum point u is obtained through the minimization sequence method in the Sobolev space $W^{1,q}(\Omega)$. The minimum point is corresponding to the generalized solution of the E–L equation. From the view point of differential equations, it is still necessary to answer whether this generalized solution has sufficient differentiability to satisfy the corresponding E–L differential equation. That is to say, can we derive $u \in C^2$ from the generalized solution u which is obtained? This type of problem is called a regularity problem. We present the following theorem, whose proof can be found in other books [7].

Theorem 3.5. *When* $1 < r < \infty$, *let* L *satisfy the growth condition as follows:*

$$|L(t, u, p)| + |L_u(t, u, p)| + |L_p(t, u, p)| \leq C(1 + |p|^r),$$

and for $r = \infty$, *no growth condition need to be added to* L.

Let the matrix $(L_{p_i p_j}(t, u, p))$, $\forall (t, u, p) \in \bar{J} \times \mathbb{R}^N \times \mathbb{R}^N$ *be positive definite, where* $J \subset \mathbb{R}$ *is an open interval. Suppose that* $u^* \in W^{1,r}(J, \mathbb{R}^N)$ *is a minimizer of the functional* $I(u) = \int_J L(t, u(t), \dot{u}(t))dt$. *Then, after changing the values of* u^* *on a zero measure set,* $u^* \in C^2$.

In variational calculus, for a functional I on space $W^{1,r}(\Omega, \mathbb{R}^N)$, the difference between the variational $\delta I(u^*, \phi)$ and the Gâteaux derivative $DI(u^*, \phi)$ is that $\phi \in C_0^1(\Omega, \mathbb{R}^N)$ or $C_0^\infty(\Omega, \mathbb{R}^N)$ in variational, and $\phi \in W_0^{1,r}(\Omega, \mathbb{R}^N)$ in Gâteaux derivative.

Example 3.1. Let $e \in C[0, T]$ be a T-periodic function with zero mean $\int_0^T e(t)dt = 0$, and a be a constant. Prove that the equation

$$\ddot{u}(t) + a \sin u(t) = e(t) \tag{3.4}$$

has a periodic solution with the period T.

Proof. Denote by $H^1_{\text{per}}(0, T)$ the subspace of the Sobolev space $H^1(0, T)$ with T-period, that is, the closure of the T-periodic C^∞ functions in $H^1(0, T)$. Define the functional

$$I(u) = \int_0^T \left(\frac{1}{2}|u'(t)|^2 + a \cos u(t) - E(t)u'(t) \right) dt,$$

where

$$E(t) = \int_0^t e(s)ds.$$

E is also a periodic function with T-period. The E–L equation of I is the equation (3.4).

For any $u \in H^1_{\text{per}}(0, T)$, we make a decomposition as

$$u = \tilde{u} + \bar{u},$$

where $\bar{u} = \frac{1}{T} \int_0^T u(t)dt$ is a real number. From the Wirtinger inequality, it can be seen that $\bar{u}$ is not controlled by the value of functional I. In other words, if the H^1-norm is directly applied to the space $H^1_{\text{per}}(0, T)$, then I is not coercive.

Note that the nonlinear term $\cos u$ in the functional I is 2π-periodic. So,

$$I(u + 2\pi) = I(u).$$

This indicates that it is not necessary to consider I on the entire $H^1_{\text{per}}(0, T)$, but rather to take a set

$$M = \{u = \xi + \eta;\ \xi \in H^1_{\text{per}}(0, T), \bar{\xi} = 0, \eta \in [0, 2\pi]\},$$

and I is restricted on M. The advantage of this is that $\bar{u}$ only varies within the bounded interval $[0, 2\pi]$.

Now, the set M is sequentially weakly closed. It is easily seen that

$$I(u) \geq \frac{1}{2}||\dot{\xi}||_2^2 - \sqrt{T}||E||_\infty||\dot{\xi}||_2 - T|a|,$$

According to the Wirtinger inequality, $||\dot{\xi}||_2$ is a norm on M, which is equivalent to the H^1 norm. I is coercive on M, and is also weakly lower semi-continuous. Hence, there is a minimizer $u \in M \subset H^1_{\text{per}}$. The conditions of the regularity theorem are also satisfied. Therefore, $u \in C^2$. This is the end.

Lemma 3.2 (Wirtinger Inequality). *Let $u \in H^1_{\text{per}}(0, T)$ and*

$$\bar{u} = \frac{1}{T}\int_0^T u(t)dt = 0.$$

Then

$$\int_0^T |u'(t)|^2 dt \geq \frac{4\pi^2}{T^2}\int_0^T |u(t)|^2 dt.$$

Proof. We expand the periodic function u on $[0, T]$ by using the Fourier series. Since $\bar{u} = 0$, then

$$u(t) = \sum_{k=1}^{\infty}\left(a_k \cos\frac{2\pi kt}{T} + b_k \sin\frac{2\pi kt}{T}\right),$$

and

$$u'(t) = \frac{2\pi}{T}\sum_{k=1}^{\infty}\left(-ka_k \sin\frac{2\pi kt}{T} + kb_k \cos\frac{2\pi kt}{T}\right).$$

By the Parseval equality, we have

$$\begin{aligned}
\int_0^T |u'(t)|^2 dt &= \frac{(2\pi)^2}{T^2}\sum_{k=1}^{\infty} k^2(|a_k|^2 + |b_k|^2) \\
&\geq \frac{(2\pi)^2}{T^2}\sum_{k=1}^{\infty}(|a_k|^2 + |b_k|^2) \\
&= \frac{(2\pi)^2}{T^2}\int_0^T |u(t)|^2 dt.
\end{aligned}$$

This is the end.

3.1.4 Conditional Extremum

Definition 3.1.2. Let X be a Banach space, and $g_i \in C^1(X, \mathbb{R})$, $i = 1, 2, \ldots, n$. The **constraint condition** $M = \{x \in X;\ g_i(x) = 0,\ i = 1, 2, \ldots, n\}$ is said to be **regular**, if $M \neq \emptyset$, and for any $x \in M$, $g_1'(x), \ldots, g_n'(x)$ are linearly independent.

Let $M = \{x \in X;\ g_i(x) = 0,\ i = 1, 2, \ldots, n\}$ be regular constraint, and $G(x) = (g_1(x), \ldots, g_n(x))$. Then $G \in C^1(X, \mathbb{R}^n)$, and for any fixed $x_0 \in M$, $G'(x_0) : X \to \mathbb{R}^n$ is surjective. Thus, X has the following direct decomposition

$$X = X_1 \oplus X_2, \quad X_1 = \text{Ker } G'(x_0).$$

For any $x - x_0 \in X$, there exists a unique $x_1 \in X_1, x_2 \in X_2$, such that

$$x - x_0 = x_1 + x_2.$$

By the implicit function theorem, there exist $\delta, r > 0$, such that when $||x_1|| \leq \delta$, there exists a unique $x_2(x_1)$, $||x_2(x_1)|| < r$, such that

$$G(x_0 + x_1 + x_2(x_1)) = 0,$$

and $x_2(x_1)$ is continuously differentiable with respect to x_1.

Let $h \in X_1$ with $||h|| = 1$. Then, when $|t| < \delta$, we have

$$G(x_0 + th + x_2(th)) = 0.$$

If we calculate the differential with respect to t and let $t = 0$, we obtain

$$G'(x_0)(h + x_2'(0)h) = 0.$$

Since $G'(x_0)h = 0$ and $G'(x_0) : X_2 \to \mathbb{R}^n$ is isomorphic, it is known that

$$x_2'(0)h = 0.$$

The above discussion indicates that if M is a regular constraint, then for any $x_0 \in M$, as well as for any $h \in \bigcap_{i=1}^n \text{Ker } g_i'(x_0)$, there exists a C^1 curve $\phi : [0, 1] \to M$ on M, such that

$$\phi(0) = x_0, \quad \phi'(0) = h.$$

From the view point of differential manifolds, this fact is evident. Because M is a C^1 manifold, then h is an element of tangent space of M at x_0. Therefore, there is a C^1 curve ϕ, such that $\phi(0) = x_0$, $\phi'(0) = h$.

Theorem 3.6. *Let X be a Banach space, $f, g_i \in C^1(X, \mathbb{R})$, $i = 1, 2, \ldots, n$. Suppose that $M = \{x \in X;\ g_i(x) = 0,\ i = 1, 2, \ldots, n\}$ is a regular constraint. If f is regarded as a real function on M and reaches a local minimum at $x_0 \in M$, then there must be n constants $\lambda_1, \ldots, \lambda_n$, such that*

$$f'(x_0) = \sum_{i=1}^{n} \lambda_i g_i'(x_0).$$

Real numbers $\lambda_1, \ldots, \lambda_n$ are called **Lagrangian multipliers**.

Proof. Since f is regarded as a real function on M and reaches a local minimum at $x_0 \in M$, then there exists $\delta > 0$, such that when $x \in M$ and $||x - x_0|| \le \delta$, we have

$$f(x) - f(x_0) \ge 0. \tag{3.5}$$

We claim that $f'(x_0)h = 0$ for any $h \in \bigcap_{i=1}^{n} \operatorname{Ker} g_i'(x_0)$. In fact, for any fixed $h \in \bigcap_{i=1}^{n} \operatorname{Ker} g_i'(x_0)$, let $\phi : [0, 1] \to M$ be a C^1 curve, such that $\phi(0) = x_0, \phi'(0) = h$. From (3.5), it follows that

$$0 \le \lim_{t \to 0^+} \frac{f(\phi(t)) - f(x_0)}{t} = f'(x_0)\phi'(0) = f'(x_0)h,$$

Replacing h by $-h$, there is also $-f'(x_0)h \ge 0$. So, $f'(x_0)h = 0$.

Take $e_1, \ldots, e_n \in X$, such that

$$g_i'(x_0)e_j = \delta_{ij} = \begin{cases} 1, & \text{if } i = j, \\ 0, & \text{if } i \ne j. \end{cases}$$

For any fixed $x \in X$, let

$$h = x - \sum_{i=1}^{n} \langle g_i'(x_0), x \rangle e_i.$$

Then $g_i'(x_0)h = 0,\ i = 1, 2, \ldots, n$. Hence,

$$0 = \langle f'(x_0), h \rangle = \langle f'(x_0), x \rangle - \sum_{i=1}^{n} \langle f'(x_0), e_i \rangle \langle g_i'(x_0), x \rangle,$$

Setting $\lambda_i = \langle f'(x_0), e_i \rangle$, then

$$\langle f'(x_0) - \sum_{i=1}^{n} \lambda_i g_i'(x_0), x \rangle = 0.$$

By the arbitrariness of x, we obtain

$$f'(x_0) = \sum_{i=1}^{n} \lambda_i g_i'(x_0).$$

This is the end.

Similar discussions can be made on constraints with inequality signs.

Definition 3.1.3. Let X be a Banach space, $g_i, h_j \in C^1(X, \mathbb{R})$, $i = 1, 2, \ldots, n, j = 1, \ldots, m$. The **constraint condition** $M = \{x \in X;\ g_i(x) = 0, h_j(x) \geq 0, i = 1, 2, \ldots, n, j = 1, \ldots, m\}$ is said to be **regular**, if $M \neq \emptyset$, and for any fixed $x_0 \in M$, $g_1'(x_0), \ldots, g_n'(x_0), h_{j_1}'(x_0), \ldots, h_{j_k}'(x_0)$ are linear independent, where $h_{j_1}, \ldots, h_{j_k}$ are all of $h_1, \ldots, h_m$ that are equal to zero at x_0.

Theorem 3.7. *Let X be a Banach space, $f, g_i, h_j \in C^1(X, \mathbb{R})$, $i = 1, 2, \ldots, n, j = 1, \ldots, m$. Suppose that $M = \{x \in X;\ g_i(x) = 0, h_j(x) \geq 0,\ i = 1, 2, \ldots, n,\ j = 1, \ldots, m\}$ is a regular constraint. If f is regarded as a real function on M and reaches a local minimum at $x_0 \in M$, then there must exist n constants $\lambda_1, \ldots, \lambda_n$ and m nonnegative real numbers $\alpha_1, \ldots, \alpha_m$, such that*

$$f'(x_0) = \sum_{i=1}^{n} \lambda_i g_i'(x_0) + \sum_{j=1}^{m} \alpha_j h_j'(x_0),$$

where $\alpha_j = 0$ whenever $h_j(x_0) > 0$.

3.1.5 Ekeland Variational Principle

The Ekeland variational principle was proposed in 1970 and is a very general and widely applicable minimization principle. It provides a method for selecting minimization sequences, which have a wide range of applications when combined with other conditions.

Theorem 3.8 (Ekeland). *Let (M, δ) be a complete metric space, and let $f : M \to \mathbb{R} \cup \{\infty\}$ be bounded from below, lower semi-continuous, and be not always equal to ∞. Given any fixed $\epsilon > 0$ and $x_\epsilon \in M$ satisfying*

$$f(x_\epsilon) < \inf_{x \in M} f(x) + \epsilon,$$

then there exists $y_\epsilon \in M$, such that

$$f(y_\epsilon) \leq f(x_\epsilon), \tag{3.6}$$

$$d(x_\epsilon, y_\epsilon) \leq 1, \tag{3.7}$$

$$f(x) > f(y_\epsilon) - \epsilon d(y_\epsilon, x), \ \forall x \in M, \ x \neq y_\epsilon. \tag{3.8}$$

Proof. (1) We define a sequence $\{y_n\}$ by using the following method.

Take $y_0 = x_\epsilon$. If y_0 satisfies (3.8), then let $y_\epsilon = y_0$.

If y_0 does not satisfy (3.8), then

$$S_0 = \{y \in M;\ f(y) \leq f(y_0) - \epsilon d(y_0, y)\}$$

also contains other point besides y_0. Take $y_1 \in S_0$ such that

$$f(y_1) \leq \frac{1}{2}(f(y_0) + \inf_{y \in S_0} f(y)) < f(y_0).$$

Generally, suppose that y_n has already been taken out. Then, if y_n satisfies (3.8), let $y_\epsilon = y_n$. In fact, from the definition of y_k, $0 \leq k \leq n$, it follows that

$$\epsilon d(y_n, y_0) \leq \epsilon \sum_{k=0}^{n-1} d(y_{k+1}, y_k) \leq f(y_0) - f(y_n) < \epsilon.$$

Then y_n satisfies (3.7), while (3.6) is clearly satisfied.

If y_n does not satisfies (3.8), then the set

$$S_n = \{y \in M;\ f(y) \leq f(y_n) - \epsilon d(y_n, y)\}$$

also contains other point besides y_n. Take $y_{n+1} \in S_n$ such that

$$f(y_{n+1}) \leq \frac{1}{2}(f(y_n) + \inf_{y \in S_n} f(y)) < f(y_n). \tag{3.9}$$

Without loss of generality, assuming this process continues, a sequence of points $\{y_n\}$ is obtained.

(2) Since

$$\begin{aligned}\epsilon d(y_n, y_{n+p}) &\le \sum_{k=0}^{p-1} \epsilon d(y_{n+k}, y_{n+k+1}) \\ &\le \sum_{k=0}^{p-1} (f(y_{n+k}) - f(y_{n+k+1})) \\ &= f(y_n) - f(y_{n+p}) \end{aligned} \tag{3.10}$$

and $f(y_n)$ is decreasing and bounded from below, it follows that $\{y_n\}$ is a Cauchy sequence.

(3) Let $y_\epsilon = \lim_{n\to\infty} y_n$. In the sequel, we just need to prove that y_ϵ satisfies (3.6), (3.7), and (3.8).

Firstly, since f is lower semi-continuous, it follows that

$$f(y_\epsilon) \le \lim_{n\to\infty} f(y_n) \le f(y_0) = f(x_\epsilon).$$

Then (3.6) is satisfied.

(3.7) follows from the following inequality

$$\begin{aligned}\epsilon d(x_\epsilon, y_\epsilon) &= \epsilon \lim_{n\to\infty} d(x_\epsilon, y_n) \\ &\le \lim_{n\to\infty} (f(x_\epsilon) - f(y_n)) \\ &\le f(x_\epsilon) - \inf_{y\in M} f(x) \le \epsilon.\end{aligned}$$

Finally, (3.8) will be verified. By contradiction, suppose that there exists some $x \neq y_\epsilon$ such that

$$f(x) \le f(y_\epsilon) - \epsilon d(y_\epsilon, x). \tag{3.11}$$

Letting $p \to \infty$ in (3.10), we have

$$\epsilon d(y_n, y_\epsilon) \le f(y_n) - f(y_\epsilon).$$

By (3.11), we obtain

$$f(x) \le f(y_n) - \epsilon(d(y_\epsilon, x) + d(y_n, y_\epsilon)) \le f(y_n) - \epsilon d(x, y_n).$$

This indicates that $x \in S_n$ hold for any n. On the other hand, from (3.9), it follows that

$$2f(y_{n+1}) - f(y_n) \le \inf_{y\in S_n} f(y) \le f(x).$$

Letting $n \to \infty$, we arrive at

$$f(y_\epsilon) \leq \lim_{n\to\infty} f(y_n) \leq f(x),$$

which contradicts to (3.11). This is the end.

The Ekeland variational principle does not obtain the minimum value of the functional f. Its significance lies in the fact that this specially selected minimization sequence provides a special set of approximate minimizer.

Corollary 3.1 (Approximate Extremum Point). *Let X be a Banach space, and $f \in C^1(X, \mathbb{R})$ be bounded from below. Then for any $\epsilon > 0$, there exists $y_\epsilon \in X$ such that*

$$f(y_\epsilon) \leq \inf_{x\in X} f(x) + \epsilon, \tag{3.12}$$

and

$$||f'(y_\epsilon)|| \leq \epsilon. \tag{3.13}$$

Proof. From Theorem 3.8, it is known that for any $\epsilon > 0$, there exists $y_\epsilon \in X$ such that (3.12) holds, and for any $x \in X$, we have

$$f(x) - f(y_\epsilon) \geq -\epsilon||x - y_\epsilon||.$$

In particular, for any fixed $h \in X$ with $||h|| = 1$, take $x = y_\epsilon + th$, $t > 0$, and then

$$\frac{f(y_\epsilon + th) - f(y_\epsilon)}{t} \geq -\epsilon.$$

Letting $t \to 0$, we obtain

$$\langle f'(y_\epsilon), h\rangle \geq -\epsilon. \tag{3.14}$$

By replacing h by $-h$, we also have

$$\langle f'(y_\epsilon), -h\rangle \geq -\epsilon,$$

so,

$$\langle f'(y_\epsilon), h\rangle \leq \epsilon.$$

Combined with (3.14), we arrive at

$$|\langle f'(y_\epsilon), h\rangle| \leq \epsilon.$$

From the arbitrariness of h, it can be inferred that

$$||f'(y_\epsilon)|| \leq \epsilon.$$

This is the end.

Definition 3.1.4 (Palais–Smale Condition). Let X be a Banach space. $f \in C^1(X, \mathbb{R})$ is said to satisfy **P. S. condition**, if any sequence $\{x_n\} \subset X$ such that

$$\{f(x_n)\} \text{ is bounded}, \quad f'(x_n) \to 0$$

has a convergent subsequence.

$f \in C^1(X, \mathbb{R})$ is said to satisfy **P. S. condition** on $f^{-1}((a, b))$, if any sequence $\{x_n\} \subset X$ such that

$$\{f(x_n)\} \subset (a, b) \text{ is bounded}, \quad f'(x_n) \to 0$$

has a convergent subsequence.

There are also some variety forms of P.S. conditions that can be conveniently applied in specific problems. H. Brezis, J. M. Coron, and L. Nirenberg said that the functional f satisfies the $(\text{P.S.})_c$ condition, if any sequence $\{x_n\} \subset X$ such that

$$f(x_n) \to c, \quad f'(x_n) \to 0$$

has a convergent subsequence. For more information, please refer to [6].

Corollary 3.2. *Let X be a Banach space, and $f \in C^1(X, \mathbb{R})$ be bounded from below and satisfy P.S. condition. Then f reaches a minimum, that is, there exists $x_0 \in X$, such that $f(x_0) = \inf_{x \in X} f(x)$.*

Proof. Take $\epsilon_n \to 0$. Then, from Corollary 3.1, it follows that there exists $\{x_n\} \subset X$ such that $f(x_n) \to \inf_{x \in X} f(x)$, $\quad f'(x_n) \to 0$. By the P.S. condition, $\{x_n\}$ has a convergent subsequence $\{x_{n_k}\}$. Let $x_0 = \lim_{k \to \infty} x_{n_k}$. Thus, $f(x_0) = \inf_{x \in X} f(x)$, and $f'(x_0) = 0$. This is the end.

Combined Ekeland variational principle with the Palais–Smale condition, Corollary 3.2 provides another criterion for the existence of minima.

3.1.6 Nehari Technique

There are many functionals that are unbounded from above or below. Seemingly, extremum methods cannot be used to find critical points.

But for some special problems, Nehari used a special technique to transform finding critical points into finding extremum problems.

Let H be a Hilbert space with inner product $\langle\cdot,\cdot\rangle$. Consider a functional $I \in C^2(H,\mathbb{R})$. We want to find the critical point of I, which is the point such that $I'(u)=0$.

Denote $G(u)=\langle I'(u),u\rangle$. Then all critical points u of I will satisfy

$$G(u)=0.$$

If the set $M=\{u\in H;\ G(u)=0\}$ is a manifold, for example, $G'(u)\neq 0$, $\forall u\in M$, then I restricted on M yields a new functional $\tilde{I}$. If $\tilde{I}$ has an extremum, then we will find the extremum of $\tilde{I}$, that is, find the conditional extremum of I.

It is natural to ask the question what is the Lagrange multipliers generated by the constraint $G(u)=0$. In fact, since

$$\tilde{I}'(u)=I'(u)-\frac{\langle I'(u),u\rangle}{||G'(u)||^2}G'(u),$$

then, $\langle I'(u),u\rangle = G(u) = 0$ on M, and as long as $\langle G'(u),u\rangle \neq 0$, there is

$$\tilde{I}'(u)=0 \Longleftrightarrow I'(u)=0.$$

3.2 Minimax Principle

In this section, we will introduce how to find other types of critical points besides the minimum (maximum) point. The basic starting point of this theory is to show the existence of critical points through the changes of the topological structure in functional level sets.

3.2.1 Pseudo-Gradient Vector Fields and Deformation Lemma

Let X be a Banach space, and $f: X\to\mathbb{R}$ be a C^1 functional. $x_0\in X$ is said to be the **critical point** of f, if $f'(x_0)=0$. The set of all critical points of f is called the **critical set**, denoted as $K=\{x\in X;\ f'(x)=0\}$. The points in $\tilde{X}=X\setminus K$ are called the **regular points** of f.

$c \in \mathbb{R}$ is said to be the **critical value** of f, if there is an $x \in K$ such that $f(x) = c$. The real numbers that are not the critical values are called the **regular values** of f.

Denote the sets

$$K_c = \{x \in K;\ f(x) = c\},$$
$$f_c = \{x \in X;\ f(x) \le c\}.$$

f_c is called a **level set**.

Lemma 3.3. *Suppose that f satisfies P.S. condition. Then, for any $a, b \in \mathbb{R}, a < b$, if $f^{-1}([a,b]) \cap K = \emptyset$, there exist $\epsilon_0, \delta_0 > 0$ such that*

$$||f'(x)|| \ge \epsilon_0, \quad \forall x \in f^{-1}([a - \delta_0, b + \delta_0]).$$

Proof. By contradiction, we assume that there exist $x_n \in f^{-1}([a - \frac{1}{n}, b + \frac{1}{n}])$, $n = 1, 2, \ldots$, such that $f'(x_n) \to 0$ (as $n \to \infty$). By the P.S. condition, there exists a convergent subsequence $x_{n_k} \to x^*$. Then, it follows that $f'(x^*) = 0$ and $x^* \in f^{-1}([a, b])$, which is a contradiction. This is the end.

Lemma 3.4. *Suppose that f satisfies the P.S. condition. Then, for any $c \in \mathbb{R}$, K_c is a compact subset.*

Proof. For any $\{x_n\} \subset K_c$, then $f(x_n) = c, f'(x_n) = 0$. By the P.S. condition, there exists a convergent subsequence $x_{n_k} \to x^*$. It is clear to see $x^* \in K_c$. Hence, K_c is a compact subset. This is the end.

The purpose of this section is to discuss how the change of the topological structures of the level sets f_c, $c \in \mathbb{R}$ depends on c. Particularly, it is shown that if $f^{-1}([a,b]) \cap K = \emptyset$, that is, if there are no critical values of f between a, b, f_a is a deformation retraction of f_b. The idea is to define a solution flow $\eta(t)$ through a pseudo-gradient vector field, such that $f(\eta(t))$ is decreasing.

Definition 3.2.1. A vector field $V : \tilde{X} \to X$ is called a **pseudo-gradient vector field** of f, if V is local Lipschitz, and satisfies the following conditions:

(1) $||V(x)|| \le 2||f'(x)||$;
(2) $\langle f'(x), V(x) \rangle \ge ||f'(x)||^2$,

where $\langle \cdot, \cdot \rangle$ represents the dual action between X^* and X.

There may be more than one pseudo-gradient vector field for f. If X is a Hilbert space and f' is local Lipschitz, we can take $V(x) = f'(x)$. At this point, the pseudo-gradient vector field is the gradient vector field.

Theorem 3.9 (Palais). *Let $f \in C^1(X, \mathbb{R})$. Then there exists a pseudo-gradient vector field of f.*

Proof. For any fixed $x_0 \in \tilde{X}$, since $f'(x_0) \neq 0$, it follows that there exists $w \in X$ with $||w|| = 1$, such that

$$\langle f'(x_0), w\rangle > \frac{2}{3}||f'(x_0)||.$$

Let

$$v = \frac{3}{2}||f'(x_0)||w.$$

Then

$$\begin{cases} ||v|| < 2||f'(x_0)||, \\ \langle f'(x_0), v\rangle > ||f'(x_0)||^2. \end{cases}$$

Due to the continuity of f', there exists a neighborhood N_{x_0} of x_0, such that when $x \in N_{x_0}$, we have

$$\begin{cases} ||v|| < 2||f'(x)||, \\ \langle f'(x), v\rangle > ||f'(x)||^2. \end{cases} \tag{3.15}$$

Since $\tilde{X}$ is metrizable, it is a paracompact space [14]. Because $\{N_{x_0};\ x_0 \in \tilde{X}\}$ is an open covering of $\tilde{X}$, there exists a locally finite refinement covering $\{N_\alpha;\ \alpha \in \Lambda\}$, where Λ is an index set. Denote

$$\rho_\alpha(x) = \mathrm{dist}(x, X \backslash N_\alpha), \quad \alpha \in \Lambda,$$

Then ρ_α is local Lipschitz. Let

$$\beta_\alpha(x) = \frac{\rho_\alpha(x)}{\sum_{\alpha' \in \Lambda} \rho_{\alpha'}(x)}, \quad \alpha \in \Lambda.$$

Since $\{N_\alpha;\ \alpha \in \Lambda\}$ is a locally finite covering of $\tilde{X}$, then the series in the denominator $\sum_{\alpha \in \Lambda} \rho_\alpha(x) > 0$ and is a finite summation. Setting

$$V(x) = \sum_{\alpha \in \Lambda} \beta_\alpha(x) v_{x_\alpha},$$

then

$$||V(x)|| \leq \sum_{\alpha\in\Lambda} \beta_\alpha(x)||v_{x_\alpha}|| \leq 2||f'(x)||,$$

$$\langle f'(x), V(x)\rangle = \sum_{\alpha\in\Lambda} \beta_\alpha(x)\langle f'(x), v_{x_\alpha}\rangle \geq ||f'(x)||^2,$$

where v_{x_α} is the vector in X, such that when $x \in N_{x_\alpha}$ and v is replaced by v_{x_α}, (3.15) holds. Therefore, V is a pseudo-gradient vector field of f. This is the end.

Theorem 3.10 (Deformation Lemma). *Let X be a Banach space and $f \in C^1(X, \mathbb{R})$ satisfy P.S. condition on $f^{-1}((a,b))$. For any $c \in (a,b)$ and any positive $\epsilon_0 > 0$ with $[c-\epsilon_0, c+\epsilon_0] \subset (a,b)$, and for any neighborhood N of K_c, then there exist $\bar{\epsilon} > \epsilon > 0, \bar{\epsilon} \leq \epsilon_0$, and a homeomorphism mapping $\eta : X \to X$ such that*

(1) $\eta(f_{c+\epsilon} \setminus N) \subset f_{c-\epsilon}$;
(2) *when* $K_c = \emptyset$, $\eta(f_{c+\epsilon}) \subset f_{c-\epsilon}$;
(3) *when* $x \notin f^{-1}([c-\bar{\epsilon}, c+\bar{\epsilon}])$, $\eta(x) = x$.

Proof. Since K_c is compact and N is a neighborhood of K_c, there exists $\delta > 0$, such that $N_\delta = \{x \in X;\ \mathrm{dist}(x, K_c) < \delta\} \subset N$. By P.S. condition, there exist $b_0, \bar{\epsilon} > 0$ such that

$$||f'(x)|| \geq b_0, \quad \forall x \in f^{-1}([c-\bar{\epsilon}, c+\bar{\epsilon}])\backslash N_{\frac{\delta}{8}}.$$

Choose a local Lipschitz functional as

$$\tilde{\chi}(x) = \begin{cases} 1, & x \notin \bar{N}_{\frac{\delta}{4}}, \\ 0, & x \in \bar{N}_{\frac{\delta}{8}}, \end{cases} \quad 0 \leq \tilde{\chi}(x) \leq 1.$$

Such a functional can always be constructed. That is because for any two disjoint closed subsets A, B in X, letting

$$g(x) = \frac{\mathrm{dist}(x, A)}{\mathrm{dist}(x, A) + \mathrm{dist}(x, B)},$$

then $g(x) = 0$ as $x \in A$, $g(x) = 1$ as $x \in B$; and $0 \leq g(x) \leq 1$.

Take $0 < \epsilon < \min\{\bar{\epsilon}, \frac{b_0\delta}{8}, \frac{b_0}{8}\}$, and choose a local Lipschitz functional as

$$\chi(x) = \begin{cases} 0, & \text{if } x \notin f^{-1}([c-\bar{\epsilon}, c+\bar{\epsilon}]), \\ 1, & \text{if } x \in f^{-1}([c-\epsilon, c+\epsilon]), \end{cases} \quad 0 \le \chi(x) \le 1.$$

Put

$$\Phi(x) = \begin{cases} -\chi(x)\tilde{\chi}(x)\frac{V(x)}{||V(x)||}, & \text{if } x \in \tilde{X}, \\ 0, & \text{if } x \in X \setminus \tilde{X}, \end{cases}$$

where V is a pseudo-gradient vector field of f. Since $f^{-1}([c-\bar{\epsilon}, c+\bar{\epsilon}]) \setminus N_{\frac{\delta}{8}} \subset \tilde{X}$, then Φ is well defined on X, $||\Phi(x)|| \le 1$, and local Lipschitz.

By the fundamental theory of differential equations, the initial value problem

$$\begin{cases} \frac{d\eta}{dt} = \Phi(\eta), \\ \eta(0) = x \end{cases}$$

admits a unique solution, denoted as $\eta(t,x)$, and the maximal interval of existence to the right of the solution is denoted by $[0,\infty)$.

From the structure of Φ, it follows that

$$\eta(t,\cdot) = id, \quad \text{as } x \notin f^{-1}([c-\bar{\epsilon}, c+\bar{\epsilon}]).$$

According to the existence and uniqueness theorem of the solution, we obtain that $\eta(0,\cdot) = id$, and $\eta(t,\cdot) : X \to X$ is homeomorphism. In the sequel, it will be verified that

$$\eta(1, f_{c+\epsilon} \setminus N_\delta) \subset f_{c-\epsilon}.$$

Firstly, we claim that (a) $||\eta(t,x)-x|| \le t$; (b) if $\eta(s,x) \in f^{-1}([c-\epsilon, c+\epsilon]) \setminus \bar{N}_{\frac{\delta}{4}}, \forall s \in [0,t)$, then $f(x) - f(\eta(t,x)) \ge \frac{b_0}{2}t$.

In fact, (a) follows from the inequality

$$||\eta(t,x) - x|| \le \int_0^t ||\eta'(s,x)||ds \le t.$$

And, (b) is due to

$$
\begin{aligned}
f(x)-f(\eta(t,x)) &= -\int_0^t \frac{d}{ds} f(\eta(s,x))ds \\
&= -\int_0^t \langle f'(\eta(s,x)), \eta'(s,x)\rangle ds \\
&= \int_0^t \chi(q)\tilde{\chi}(q)\langle f'(q), \frac{V(q)}{||V(q)||}\rangle\Big|_{q=\eta(s,x)} ds \\
&\geq \frac{b_0}{2}t.
\end{aligned}
$$

We will complete the proof in two steps.

(1) Show that for any $x \in f^{-1}([c-\epsilon, c+\epsilon]) \setminus N_\delta$, there exists $t_x \leq \frac{4}{b_0}\epsilon$, such that $\eta(t_x, x) \in f_{c-\epsilon} \cup \bar{N}_{\frac{\delta}{4}}$.

By contradiction, assume that there is some $x \in f^{-1}([c-\epsilon, c+\epsilon]) \setminus N_\delta$, such that $\eta(t,x) \notin f_{c-\epsilon} \cup \bar{N}_{\frac{\delta}{4}}$ for any $t \in [0, \frac{4}{b_0}\epsilon]$. Then, $\eta(t,x) \in f^{-1}([c-\epsilon, c+\epsilon]) \setminus N_{\frac{\delta}{4}}$, $t \in [0, \frac{4}{b_0}\epsilon]$. By (b), we have

$$
2\epsilon > f(x) - f\left(\eta\left(\frac{4}{b_0}\epsilon, x\right)\right) \geq \frac{b_0}{2} \times \frac{4}{b_0}\epsilon = 2\epsilon,
$$

which is a contradiction.

(2) Show that for any $x \in f^{-1}([c-\epsilon, c+\epsilon]) \setminus N_\delta$, if there exists $t_x \leq \frac{4}{b_0}\epsilon$ such that $\eta(t_x, x) \in \bar{N}_{\frac{\delta}{4}}$, then $\eta(t_x, x) \in f_{c-\epsilon}$.

Indeed, letting

$$
t_1 = \sup\{t \geq 0;\ \eta(\tau, x) \notin N_{\frac{\delta}{4}},\ \forall \tau \in [0,t)\},
$$

then $t_1 \leq t_x$, $\eta(\tau, x) \notin N_{\frac{\delta}{4}}$ as $\tau \in [0, t_1)$, and $\eta(t_1, x) \in \bar{N}_{\frac{\delta}{4}}$. Since the functional f is decreasing along the solution flow, i.e.,

$$
f(x) - f(\eta(t,x)) = \int_0^t \chi(q)\tilde{\chi}(q)\langle f'(q), \frac{V(q)}{||V(q)||}\rangle\Big|_{q=\eta(s,x)} ds \geq 0,
$$

it suffices to show $\eta(t_1, x) \in f_{c-\epsilon}$. By contradiction, assume $\eta(t_1, x) \notin f_{c-\epsilon}$, then $\eta(t,x) \in f^{-1}([c-\epsilon, c+\epsilon])$, $\forall t \in [0, t_1]$. From the claim (a) and (b), it follows that

$$\begin{aligned}||\eta(t_1, x) - x|| \le t_1 &\le \frac{2}{b_0}(f(x) - f(\eta(t_1, x))) \\ &\le \frac{4}{b_0}\epsilon < \frac{\delta}{2}.\end{aligned}$$

Since $x \notin N_\delta$, thus $\eta(t_1, x) \notin \bar{N}_{\frac{\delta}{2}}$, which is a contradiction.

By the above two steps and $t_1 < 1$, as long as $\eta(\cdot) = \eta(1, \cdot)$ is taken, the conclusion holds. This is the end.

3.2.2 Minimax Principle

Theorem 3.11 (Minimax Principle). *Let X be a Banach space, $f \in C^1(X, \mathbb{R})$, and $\mathfrak{F}$ be a subset family that is consisted of some subsets of X. Denote*

$$c = \inf_{F \in \mathfrak{F}} \sup_{x \in F} f(x).$$

Suppose that

(1) *c is a finite number;*
(2) *there exists $\epsilon_0 > 0$, such that for any continuous mapping $\eta : X \to X$ satisfying $\eta\big|_{f_{c-\epsilon_0}} = id\big|_{f_{c-\epsilon_0}}$, and for any $F \in \mathfrak{F}$, there is $\eta(F) \in \mathfrak{F}$;*
(3) *f satisfies the P.S. condition on $f^{-1}((c - \epsilon_0, c + \epsilon_0))$.*

Then c is a critical value of f.

Proof. By contradiction, assume that c is no critical value of f, *i.e.*, $K_c = \emptyset$. By the deformation lemma, there exist $\bar{\epsilon} > \epsilon > 0$, $\bar{\epsilon} < \epsilon_0$, and continuous mapping $\eta : X \to X$, such that $\eta\big|_{f_{c-\bar{\epsilon}}} = id\big|_{f_{c-\bar{\epsilon}}}$, and $\eta(f_{c+\epsilon}) \subset f_{c-\epsilon}$. According to the definition of c, there exists $F_0 \in \mathfrak{F}$ such that

$$c \le \sup_{x \in F_0} f(x) < c + \epsilon.$$

Thus

$$\sup_{x\in\eta(F_0)} f(x) \le c-\epsilon.$$

From the condition (2), it follows that $\eta(F_0)\in\mathfrak{F}$. Therefore,

$$c=\inf_{F\in\mathfrak{F}}\sup_{x\in F} f(x)\le \sup_{x\in\eta(F_0)} f(x)\le c-\epsilon,$$

which is a contradiction. This is the end.

There are two main ways to choose the subset family $\mathfrak{F}$: to construct the subset family through the concept of the linking in topology; to construct the subset family through the category, index, pseudo-index.

Definition 3.2.2. Let X be a Banach space, Q, Q_0, S be closed subsets of X, $Q_0\subset Q$. (Q,Q_0) is said **linking** with S, if

(1) $Q_0\cap S=\emptyset$.
(2) For any continuous mapping $\phi: Q\to X$ satisfying $\phi\Big|_{Q_0}=id\Big|_{Q_0}$, we have

$$\phi(Q)\cap S\neq\emptyset.$$

Theorem 3.12. *Let X be a Banach space, Q, Q_0, S be closed subsets of X, $Q_0\subset Q$, and (Q,Q_0) be linking with S. Suppose that $f\in C^1(X,\mathbb{R})$ satisfies the following conditions:*

(1) $\sup_{x\in Q} f(x)<r<\infty$;
(2) *there exist real numbers α,β $(\beta>\alpha)$, such that*

$$\begin{aligned}&\sup_{x\in Q_0} f(x)\le\alpha,\\ &\inf_{x\in S} f(x)\ge\beta;\end{aligned}\tag{3.16}$$

(3) *f satisfies the P.S. condition on $f^{-1}((\alpha,r))$.*

Define the subset family of X as follows

$$\mathfrak{F}=\left\{\phi(Q);\ \phi\in C(Q,X),\phi\Big|_{Q_0}=id\Big|_{Q_0}\right\}.$$

Let

$$c = \inf_{\phi(Q)\in\mathfrak{F}} \sup_{x\in\phi(Q)} f(x).$$

Then $c \geq \beta$, and c is a critical value of f.

Proof. By the minimax principle, we only need to verify that the conditions (1) and (2) in Theorem 3.11 are satisfied.

Let $\phi = id$. Then $Q \in \mathfrak{F}$. From the condition (1), it follows that

$$c \leq \sup_{x\in Q} f(x) < r.$$

From the definition of the linking, for any $\phi(Q) \in \mathfrak{F}$, we have $\phi(Q) \cap S \neq \emptyset$. Combined with (3.16), we arrive at

$$\sup_{x\in\phi(Q)} f(x) \geq \inf_{x\in S} f(x) \geq \beta.$$

Then

$$c = \inf_{\phi(Q)\in\mathfrak{F}} \sup_{x\in\phi(Q)} f(x) \geq \beta.$$

So, $\beta \leq c < r$, that is, the condition (1) in Theorem 3.11 is satisfied.

If let $\epsilon_0 = \min\{\beta - \alpha, r - c\}$, then $Q_0 \subset f_{c-\epsilon_0}$. So, it is easily known that the condition (2) in Theorem 3.11 is satisfied. This is the end.

Theorem 3.13 (Mountain Pass Theorem). *Let X be a Banach space, and $f \in C^1(X, \mathbb{R})$. Suppose that there exist a neighborhood U of the origin point in X, $x_0 \notin U$, and a constant β, such that*

$$f(0),\ f(x_0) < \beta,$$

$$f\Big|_{\partial U} \geq \beta.$$

Let

$$c = \inf_{\phi\in\Gamma} \max_{t\in[0,1]} f(\phi(t)),$$

where $\Gamma = \{\phi \in C([0,1], X);\ \phi(0) = 0,\ \phi(1) = x_0\}$. If f satisfies the P.S. condition on $f^{-1}((c-\epsilon_0, c+\epsilon_0))$, then c is a critical value of f.

Proof. Let $Q = \{tx_0;\ t \in [0,1]\}$, $Q_0 = \{0, x_0\}$, and $S = \partial U$. Then (Q, Q_0) is linking with $S = \partial U$. Theorem 3.12 directly implies the conclusion. This is the end.

The mountain pass theorem was proposed by A. Ambrosetti and P.H. Rabinowitz in 1974. The P.S. condition plays a crucial role in the mountain pass theorem. If it does not hold, there is the following Brezis–Nirenberg counterexample.

Consider the function $f(x,y) = x^2 + (1-x)^3 y^2$ in $\mathbb{R}^2$. Let $c = \inf_{x^2+y^2=\frac{1}{4}} f(x,y) > 0$. Then f possesses a floor: $\Omega = \{(x,y) \in \mathbb{R}^2;\ f(x,y) \leq c\}$, and $f(0,0) = 0, f(4,1) = -11$. But, a direct verification show that f only has one critical point $(0,0)$.

There are two cases that the sets are linking below. Combined with Theorem 3.12, we can obtain specific minimax theorems.

Example 3.2. Let X be a Banach space, X_1 be its finite dimensional subspace, and X_2 be the complement space of X_1, i.e., $X = X_1 \oplus X_2$. Let $S = X_2$, $Q = B_R \cap X_1$, and $Q_0 = S_R \cap X_1$, where $B_R = \{x \in X;\ ||x|| \leq R\}$ and $S_R = \partial B_R = \{x \in X;\ ||x|| = R\}$. Then (Q, Q_0) is linking with S.

Proof. It is clear to see $S \cap Q_0 = \emptyset$. We only need to verify that for any continuous mapping $\phi : Q \to X$ with $\phi\big|_{Q_0} = id|_{Q_0}$, there always is

$$\phi(Q) \cap S \neq \emptyset,$$

which is equivalent to proving that there exists $x_0 \in Q$, such that

$$P\phi(x_0) = 0,$$

where $P : X \to X_1$ is the projection operator. For this purpose, we denote

$$\begin{aligned} &H : Q \times [0,1] \to X_1, \\ &H(x,t) = tP\phi(x) + (1-t)x. \end{aligned}$$

When $t \in [0,1]$ and $x \in \partial Q = Q_0$, there is

$$H(x,t) = x \neq 0.$$

By the homotopy invariance and normality, we arrive at

$$\deg(P\phi, Q, 0) = \deg(id, Q, 0) = 1.$$

By the Kronecker existence theorem, we obtain that there exists $x_0 \in Q$, such that $P\phi(x_0) = 0$.

Example 3.3. Let X be a Banach space, X_1 be its finite dimensional subspace, X_2 be the complement space of X_1, and $e \in X_2$ with $||e|| = 1$. Let r_1, r_2, ρ be positive numbers, and $r_1 > \rho$. Denote

$$\begin{aligned} S &= X_2 \cap \partial B_\rho, \\ Q &= \{u + te;\ u \in X_1 \cap B_{r_2}, t \in [0, r_1]\}, \\ Q_0 &= \partial Q, \end{aligned}$$

where B_r is the ball with radius r centered at the origin point. Then (Q, Q_0) is linking with S.

Proof. It is clear to see $S \cap Q_0 = \emptyset$. In the sequel, we want to verify that for any continuous mapping $\phi : Q \to X$ with $\phi\Big|_{Q_0} = id\Big|_{Q_0}$, there always is $\phi(Q) \cap S \neq \emptyset$, that is, to verify that there exists $x_0 + s_0 e \in Q$, such that

$$\begin{cases} P\phi(x_0 + s_0 e) = 0, \\ ||\phi(x_0 + s_0 e)|| = \rho, \end{cases}$$

where $P : X \to X_1$ is the projection.

Make the homotopy as follows:

$$\begin{aligned} H :\ & Q \times [0, 1] \to X_1 \times \mathrm{span}\{e\}, \\ H(x + se, t) &= [(1 - t)x + tP\phi(x + se)] \\ &\quad + [(1 - t)s + t||(I - P)\phi(x + se)|| - \rho]e. \end{aligned}$$

Then, when $t \in [0, 1]$ and $x + se \in \partial Q = Q_0$, we have

$$H(x + se, t) = x + (s - \rho)e \neq 0.$$

From the homotopy invariance of Brouwer degree, it follows that

$$\deg(H(\cdot, 1), Q, 0) = \deg(H(\cdot, 0), Q, 0).$$

At this moment, $H(x+se,0) = x+(s-\rho)e$. By the reduction theorem of Brouwer degree, we arrive at

$$\deg(H(\cdot,0),Q,0) = \deg(s-\rho,(0,r_1),0) = 1.$$

Thus,

$$\deg(H(\cdot,1),Q,0) = 1.$$

So, there exists $x_0 + s_0 e \in Q$, such that $H(x_0 + s_0 e, 1) = 0$. Since

$$H(x_0 + s_0 e, 1) = P\phi(x_0 + s_0 e) + (||(I-P)\phi(x_0 + s_0 e)|| - \rho)e,$$

we arrive at

$$\begin{cases} P\phi(x_0 + s_0 e) = 0, \\ ||(I-P)\phi(x_0 + s_0 e)|| - \rho = ||\phi(x_0 + s_0 e)|| - \rho = 0. \end{cases}$$

This is the end.

3.3 $\mathbb{Z}_2$ Index and Category

$\mathbb{Z}_2$ index was proposed by Krasnosel'skii and Coffman et al. in the 1950s to 1970s. Their ideas were derived from the Ljusternik–Schnirelman theory established in the 1930s.

3.3.1 $\mathbb{Z}_2$ Index

Let X be a Banach space. The set $A \subset X$ is called **symmetric with respect to the origin point**, if $x \in A$ implies that $-x \in A$. Denote $\mathcal{A}$ as the family of sets consisting of all closed symmetric subsets with respect to the origin point in X.

Definition 3.3.1. The nonnegative integer mapping $i : \mathcal{A} \to \mathbb{Z}_+ \cup \{\infty\}$ is called the $\mathbb{Z}_2$ **index** on $\mathcal{A}$ or the **genus** on $\mathcal{A}$, if $i(A)$ is defined as follows:

Case (1) $A = \emptyset$. Let $i(A) = 0$.

Case (2) $A \neq \emptyset$. If there exists an integer m and a continuous odd mapping $\phi : A \to \mathbb{R}^m \setminus \{0\}$, then $i(A)$ is defined as the smallest integer of m with this property, i.e.,

$$i(A) = \min\{m \in \mathbb{Z}_+;\ \text{there exists a continuous odd mapping}\\ \phi : A \to \mathbb{R}^m \setminus \{0\}\}.$$

Case (3) If there does not exist the continuous odd mapping mentioned above for any positive integer m, then let $i(A) = \infty$.

Theorem 3.14. *The $\mathbb{Z}_2$ index i on $\mathcal{A}$ possesses the following properties*:

(I_1) $i(A) = 0 \Longleftrightarrow A = \emptyset$.
(I_2) (**Normalization**) *If $A \in \mathcal{A}$ is only consisted of one pair of symmetric points, then $i(A) = 1$.*
(I_3) (**Monotonicity property**) *For any $A, B \in \mathcal{A}$, if $A \subset B$, then $i(A) \leq i(B)$.*
(I_4) (**Subadditivity**) $i(A \cup B) \leq i(A) + i(B)$, $\forall A, B \in \mathcal{A}$.
(I_5) (**Mapping property**) *For any continuous odd mapping $\phi : X \to X$ and any $A \in \mathcal{A}$, then $i(A) \leq i(\overline{\phi(A)})$.*
(I_6) (**Continuity property**) *For any $A \in \mathcal{A}$, if A is compact, then there exists a symmetric neighborhood N of A, such that $i(\bar{N}) = i(A)$. Furthermore, if A is compact and $0 \notin A$, then $i(A) < \infty$.*
(I_7) *Let X_1 be m dimensional subspace of X and S be the unit sphere of X. Then $i(X_1 \cap S) = m$.*
(I_8) *Let $X = X_1 \oplus X_2$ and* $\dim X_1 = m$. *If $A \in \mathcal{A}$ and $i(A) > m$, then $A \cap X_2 \neq \emptyset$.*

Remark. The definition of the index can be defined by using (I_1) – (I_6) in Theorem 3.14, that is, (I_1) – (I_6) in Theorem 3.14 can be regarded as the axiomatic definition of the index.

Proof. (I_1), (I_2), (I_3) directly follow from the definition. In the sequel, we will prove (I_4) – (I_8).

(I_4) Let $i(A) = m$, $i(B) = n$. Without loss of generalization, assume $m, n < \infty$. According to the definition, there exist continuous odd mappings $\phi : A \to \mathbb{R}^m \setminus \{0\}$, $\psi : B \to \mathbb{R}^n \setminus \{0\}$. By using the Tietze extension theorem, there exist continuous mappings $\tilde{\phi} : X \to$

$\mathbb{R}^m$, $\tilde{\psi} : X \to \mathbb{R}^n$, such that $\tilde{\phi}\Big|_A = \phi$, $\tilde{\psi}\Big|_B = \psi$. Let

$$\phi^*(x) = \frac{1}{2}(\tilde{\phi}(x) - \tilde{\phi}(-x)), \quad \psi^*(x) = \frac{1}{2}(\tilde{\psi}(x) - \tilde{\psi}(-x)).$$

Then, ϕ^*, ψ^* are odd, and $\phi^*\Big|_A = \phi$, $\psi^*\Big|_B = \psi$. Denote

$$\eta(x) = (\phi^*(x), \psi^*(x)).$$

Thus, $\eta : A \cup B \to \mathbb{R}^{m+n} \setminus \{0\}$ is continuous and satisfies $\eta(-x) = -\eta(x)$. From the definition of $i(A \cup B)$, we arrive at

$$i(A \cup B) \leq m + n = i(A) + i(B).$$

(I_5) Let $A \in \mathcal{A}$, and $\phi : X \to X$ be continuous and odd. Without loss of generalization, assume $i(\overline{\phi(A)}) = n < \infty$. Then, there exists a continuous odd mapping $\psi : \overline{\phi(A)} \to \mathbb{R}^n \setminus \{0\}$. Put $\eta = \psi \circ \phi$. Thus, $\eta : A \to \mathbb{R}^n \setminus \{0\}$ is continuous and odd. Hence, $i(A) \leq n = i(\overline{\phi(A)})$.

(I_6) Let $A \in \mathcal{A}$ be compact. Without loss of generalization, assume $i(A) = n < \infty$. Then, there exists a continuous odd mapping $\phi : A \to \mathbb{R}^n \setminus \{0\}$. By using the Tietze extension theorem, there exists a continuous mapping $\tilde{\phi} : X \to \mathbb{R}^n$ such that $\tilde{\phi}\Big|_A = \phi$. Let

$$\phi^*(x) = \frac{1}{2}(\tilde{\phi}(x) - \tilde{\phi}(-x)).$$

Thus, $\phi^* : X \to \mathbb{R}^n$ is continuous and odd, and $\phi^*\Big|_A = \phi$. Since A is compact and $0 \notin \phi^*(A)$, then there exist a positive ϵ and an ϵ neighborhood U_ϵ of $\phi^*(A)$ in $\mathbb{R}^n$, such that $\bar{U}_\epsilon$ does not contain the origin point. Let $N = \phi^{*-1}(U_\epsilon)$. Thus, N is a neighborhood of A and symmetric with respect to the origin point, and $\phi^* : \bar{N} \to \mathbb{R}^n \setminus \{0\}$ is continuous and odd. Hence, $i(\bar{N}) \leq n = i(A)$. From the monotonicity property, it follows that $i(\bar{N}) = i(A)$.

Furthermore, if A is compact and does not contain the origin point. Then, for each x in A, it follows from the normalization that $i(\{x, -x\}) = 1$. By using the above method, there exists a symmetric neighborhood N_ϵ of $\{x, -x\}$ with respect to the origin point, such that $i(\bar{N}_\epsilon) = 1$. Hence, $\{N_x;\ x \in A\}$ is an open covering of A. Since A is compact, there exist a finite points $x_1, \ldots, x_n \in A$, such that

$$A \subset \bigcup_{k=1}^{n} N_{x_k}.$$

By the monotonicity property and subadditivity, we arrive at

$$i(A) \le i\left(\bigcup_{k=1}^{n} \bar{N}_{x_k}\right) \le \sum_{k=1}^{n} i(\bar{N}_{x_k}) = n.$$

(I_7) Since $X_1 \cap S$ is an unit sphere of X_1 and $\dim X_1 = m$, then for any positive integer $n < m$ and any continuous odd mapping $\phi : X_1 \cap S \to \mathbb{R}^n$, it follows from the Borsuk theorem that ϕ must have at least one zero point. Thus, $i(X_1 \cap S) \ge m$. It is clear to see $i(X_1 \cap S) \le m$.

(I_8) By contradiction, assume that $A \cap X_2 = \emptyset$. Let $P : X \to X_1$ be the projection. Then, P is a continuous odd mapping, and $Px \ne 0$ for any $x \in A$. According to the definition, $i(A) \le \dim X_1 = m$, which is a contradiction. This is the end.

Theorem 3.15. *Let X be a Banach space, $\mathcal{A}$ be the family consisted of closed symmetric subsets in X with respect to the origin, i be the $\mathbb{Z}_2$ index on $\mathcal{A}$. Suppose that $f \in C^1(X, \mathbb{R})$ is an even functional and satisfies the P.S. condition. Put*

$$c_m = \inf_{A \in \mathcal{A}_m} \sup_{x \in A} f(x), \quad m = 1, 2, \ldots,$$

where $\mathcal{A}_m = \{A \in \mathcal{A};\ i(A) \ge m\}$. Then,

(1) $c_m \le c_{m+1} \le f(0), \quad m = 1, 2, \ldots;$
(2) *if $c_m > -\infty$, then c_m is a critical value of f;*
(3) *if $-\infty < c = c_m = c_{m+1} = \cdots = c_{m+k}$, then $i(K_c) \ge k + 1$.*

Proof. Due to the fact that the index of a single point set containing only the origin is infinite, it follows from $\mathcal{A}_{m+1} \subset \mathcal{A}_m$ that the conclusion (1) holds.

Because (2) is a special case of (3), the following only provides a proof of (3).

By contradiction, assume that $i(K_c) = r < k + 1$. Since K_c is compact, it follows from the index property (I_6) that there exists a neighborhood N of K_c such that $i(K_c) = i(\bar{N}) = r$. By using the deformation lemma, there exist $\epsilon > 0$ and homeomorphism mapping $\eta : X \to X$, such that $\eta(f_{c+\epsilon} \setminus N) \subset f_{c-\epsilon}$ (at this moment, you can also request that η is odd). For the above $\epsilon > 0$, it follows from the definition of c_{m+k} that there exists $A \in \mathcal{A}_{m+k}$ such that

$$c \leq \sup_{x \in A} f(x) < c + \epsilon.$$

Thus,

$$\sup_{x \in \eta(A \setminus N)} f(x) < c - \epsilon. \tag{3.17}$$

By using the subadditivity and mapping property of the index, it is known from $A \in \mathcal{A}_{m+k}$ that

$$m + k \leq i(A) \leq i(A \setminus N) + i(\bar{N}) \leq i(\eta(A \setminus N)) + r.$$

So, $i(\eta(A \setminus N)) \geq m$. Therefore, $\eta(A \setminus N) \in \mathcal{A}_m$. From the definition of c_m, we obtain

$$\sup_{x \in \eta(A \setminus N)} f(x) \geq c_m = c,$$

which is in contradiction with (3.17). This is the end.

When $c < f(0)$ and $i(K_c) \geq k + 1$, then K_c contains at least $k+1$ pairs of different points.

Theorem 3.16. *Let X be a Banach space, $f \in C^1(X, \mathbb{R})$ be an even functional, admit the P.S. condition, and satisfy the following conditions:*

(1) *there exist r-dimensional subspace V of X and $\rho > 0$, such that*

$$\sup_{x \in V \cap S_\rho} f(x) < f(0),$$

where $S_\rho = \{x \in X;\ ||x|| = \rho\}$;

(2) *there exists a closed subspace X_2 of X such that the complement space of X_2 is s-dimensional, and*

$$\inf_{x \in X_2} f(x) > -\infty.$$

Then, when $r > s$, f has at least $r - s$ pairs of different critical points.

Proof. Let $\mathcal{A}$ be the family consisted of closed symmetric subsets in X with respect to the origin, i be the $\mathbb{Z}_2$ index on $\mathcal{A}$. By using the property (I_8) of the $\mathbb{Z}_2$ index, we know that for any $A \in \mathcal{A}$ with $i(A) > s$, there is $A \cap X_2 \neq \emptyset$. By the condition (2), we obtain

$$\sup_{x\in A} f(x) \geq \inf_{x\in X_2} f(x) > -\infty.$$

Denote

$$c_m = \inf_{A\in\mathcal{A}_m} \sup_{x\in A} f(x), \quad m = 1, 2, \ldots,$$

where $\mathcal{A}_m = \{A \in \mathcal{A};\ i(A) \geq m\}$. Then, when $m > s$, c_m is finite. On the other hand, from the property (I_7) of the $\mathbb{Z}_2$ index, it follows that $i(V \cap S_\rho) = r$. Combined with the condition (1), we obtain

$$-\infty < c_{s+1} \leq c_{s+2} \leq \cdots \leq c_r < f(0).$$

By using Theorem 3.15, we conclude that f has at least $r - s$ pairs of different critical points. This is the end.

3.3.2 $\mathbb{Z}_2$ Pseudo-Index

For some unbounded functionals from above and below, the minimax values constructed through the $\mathbb{Z}_2$ index are likely not finite numbers. For this reason, the $\mathbb{Z}_2$ index has been generalized into $\mathbb{Z}_2$ pseudo-index, and the minimax values constructed through the $\mathbb{Z}_2$ pseudo-index are often finite.

Definition 3.3.2. Let X be a Banach space, $\mathcal{A}$ be the family consisted of closed symmetric subsets in X with respect to the origin, and i be the $\mathbb{Z}_2$ index on $\mathcal{A}$. Let H^* be the family consisted of continuous odd mappings on X, $\mathcal{A}^* \subset \mathcal{A}$, and $i^* : \mathcal{A}^* \to \mathbb{Z}_+ \cup \{\infty\}$ be nonnegative integer function. $(\mathcal{A}^*, i^*)$ is called a **pseudo-index** of i with respect to H^*, if the following conditions are satisfied:

(I_1^*) When $A \in \mathcal{A}^*$, $B \in \mathcal{A}$, and $\eta \in H^*$, $\overline{A\backslash B}, \overline{\eta(A)} \in \mathcal{A}^*$.
(I_2^*) When $A, B \in \mathcal{A}^*$, and $A \subset B$, $i^*(A) \leq i^*(B)$.

(I_3^*) For any $A \in \mathcal{A}^*$ and $B \in \mathcal{A}$, if $i(B) < \infty$, then $i^*(\overline{A \backslash B}) \geq i^*(A) - i(B)$.
(I_4^*) $i^*(\overline{\eta(A)}) \geq i^*(A), \quad \forall A \in \mathcal{A}^*, \eta \in H^*$.

There is great flexibility in the definition of $\mathbb{Z}_2$ pseudo-index. This is mainly reflected in the selection about H^* and $\mathcal{A}^*$. Usually, the selection of H^* and $\mathcal{A}^*$ should be combined with the functional f being studied.

Theorem 3.17. *Let i be the $\mathbb{Z}_2$ index on $\mathcal{A}$, and $f \in C^1(X, \mathbb{R})$ be an even functional and satisfy P.S. condition on $f^{-1}((a,b))$. Let H^* be the family consisted of some odd homeomorphism mappings on X, such that for any such odd homeomorphism mapping η, if $\eta\big|_{f^{-1}(\mathbb{R}\backslash(a,b))} = id\big|_{f^{-1}(\mathbb{R}\backslash(a,b))}$ and $f(\eta(x)) \leq f(x)$, there is $\eta \in H^*$. Suppose that $(\mathcal{A}^*, i^*)$ is the $\mathbb{Z}_2$ pseudo index of i with respect to H^*. Denote*

$$c_m^* = \inf_{A \in \mathcal{A}_m^*} \sup_{x \in A} f(x), \quad m = 1, 2, \ldots,$$

where $\mathcal{A}_m^* = \{A \in \mathcal{A};\ i^*(A) \geq m\}$. Then, when

$$c = c_m^* = c_{m+1}^* = \cdots = c_{m+k}^* \in (a, b),$$

c is a critical value of f, and $i(K_c) \geq k+1$.

Proof. By contradiction, assume that $i(K_c) = r < k+1$. Since K_c is compact, it follows from the property (I_6) of the index that there exists a neighborhood N of K_c, such that $i(K_c) = i(\bar{N}) = r$. By the deformation lemma, there exist $\epsilon > 0$ and homeomorphism mapping $\eta : X \to X$, such that $\eta(f_{c+\epsilon} \setminus N) \subset f_{c-\epsilon}$ (at this moment, you can also request that η is odd). For the above $\epsilon > 0$, it follows from the definition of c_{m+k}^* that there exists $A \in \mathcal{A}_{m+k}^*$ such that

$$c \leq \sup_{x \in A} f(x) < c + \epsilon.$$

Thus,

$$\sup_{x \in \eta(A \backslash N)} f(x) < c - \epsilon. \tag{3.18}$$

According to the properties of the pseudo-index, we have

$$m + k \leq i^*(A) \leq i^*(\overline{A \backslash N}) + i(\bar{N}) \leq i^*(\overline{\eta(A \backslash N)}) + r.$$

Thus, $i^*(\overline{\eta(A\backslash N)}) \geq m$. So, $\overline{\eta(A\backslash N)} \in \mathcal{A}_m^*$. From the definition of c_m^*, it is known that

$$\sup_{x \in \overline{\eta(A\backslash N)}} f(x) \geq c_m^* = c,$$

which is in contradiction with (3.18). This is the end.

In the sequel, we introduce how to select the pseudo-index.

Theorem 3.18. *Let i be the index on $\mathcal{A}$, H^* be a group composed of some odd homeomorphism mappings on X, $S \in \mathcal{A}$ be a given set. Denote*

$$i^*(A) = \inf_{h \in H^*} i(h(A) \cap S).$$

Then $(\mathcal{A}, i^*)$ is a pseudo-index of i with respect to H^*.

Proof. $(\mathrm{I}_1^*), (\mathrm{I}_2^*)$ are clearly satisfied. In the sequel, we only consider (I_3^*) and (I_4^*).

For any $A, B \in \mathcal{A}$ with $i(B) < \infty$, we have

$$\begin{aligned}
i^*(\overline{A\backslash B}) &= \inf_{h \in H^*} i(h(\overline{A\backslash B}) \cap S) \\
&\geq \inf_{h \in H^*} i(\overline{h(A) \cap S\backslash h(B)}) \quad \text{(monotonicity)} \\
&\geq \inf_{h \in H^*} [i(h(A) \cap S) - i(h(B))] \quad \text{(subadditivity)} \\
&\geq \inf_{h \in H^*} i(h(A) \cap S) - i(B) \quad (h \text{ is homeomorphism}) \\
&= i^*(A) - i(B).
\end{aligned}$$

Thus, (I_3^*) holds. For any $\eta \in H^*$ and $A \in \mathcal{A}$, we have

$$\begin{aligned}
i^*(\eta(A)) &= \inf_{h \in H^*} i(h \circ \eta(A) \cap S) \\
&\geq \inf_{h' \in H^*} i(h'(A) \cap S) \quad (H^* \text{ is group}) \\
&= i^*(A).
\end{aligned}$$

So, (I_4^*) holds. This is the end.

The selection of S usually depends on the functional being considered. In general, the selected S should be such that $\inf_{x\in S} f(x) = \alpha > -\infty$. In the following, we will select two concretes S and compute the pseudo-index for some closed symmetric subsets.

Theorem 3.19. *Let i be the $\mathbb{Z}_2$ index on $\mathcal{A}$, H^* be a group composed of some odd homeomorphism mappings on X, X_1 be a closed linear subspace of X, and X_2 be the complement space of X_1 and finite dimensional. Denote*

$$i_1^*(A) = \inf_{h\in H^*} i(h(A)\cap X_1).$$

Then, for any finite dimensional subspace V of X and any positive number ρ, we have

$$i_1^*(V\cap S_\rho) \geq \text{dim}V - \text{dim}X_2,$$

where $S_\rho = \{x \in X;\ ||x|| = \rho\}$.

Proof. Without loss of generalization, assume $\text{dim}V - \text{dim}X_2 > 0$. For any fixed odd homeomorphism mapping $h \in H^*$, let $i(h(V\cap S_\rho)\cap X_1) = k$. Since $h(V\cap S_\rho)\cap X_1$ is compact, from the continuity property of the $\mathbb{Z}_2$ index, it is known that there exists δ neighborhood N_δ of $h(V\cap S_\rho)\cap X_1$, such that

$$i(\bar{N}_\delta) = i(h(V\cap S_\rho)\cap X_1) = k.$$

Let $P : X \to X_2$ be projection. Then $P(h(V\cap S_\rho)\backslash N_\delta) \subset X_2 \setminus \{0\}$. Hence, $i(h(V\cap S_\rho)\setminus N_\delta) \leq \text{dim}X_2$. By the subadditivity and (I_7) of the $\mathbb{Z}_2$ index, we obtain

$$\begin{aligned}\text{dim}V &= i(V\cap S_\rho) = i(h(V\cap S_\rho))\\ &\leq i(h(V\cap S_\rho)\setminus N_\delta) + i(\bar{N}_\delta) \leq \text{dim}X_2 + k.\end{aligned}$$

So, $k \geq \text{dim}V - \text{dim}X_2$. This is the end.

Theorem 3.20. *Let i be the $\mathbb{Z}_2$ index on $\mathcal{A}$, H^* be a group composed of some bounded odd homeomorphism mappings on X, X_1 be a closed* *linear subspace of* X*, and the complement space* X_2 *of* X_1 *is finite dimensional. For* $A \in \mathcal{A}$*, denote*

$$i_2^*(A) = \inf_{h \in H^*} i(h(A) \cap S_\rho \cap X_1), \tag{3.19}$$

where $\rho > 0$ *and* $S_\rho = \{x \in X;\ ||x|| = \rho\}$*. Then, for any finite dimensional subspace* V *of* X*, there is*

$$i_2^*(V) \geq \dim V - \dim X_2.$$

Proof. Without loss of generalization, assume that $\dim V - \dim X_2 > 0$. For any fixed odd mapping $h \in H^*$, let $i(h(V) \cap S_\rho \cap X_1) = k$. Since $h(V) \cap S_\rho \cap X_1$ is compact, it is known from the continuity property of the $\mathbb{Z}_2$ index that there exists δ neighborhood N_δ of $h(V) \cap S_\rho \cap X_1$, such that

$$i(\bar{N}_\delta) = i(h(V) \cap S_\rho \cap X_1) = k.$$

Let $P : X \to X_2$ be the projection. Then, $P((h(V) \cap S_\rho) \backslash N_\delta) \subset X_2 \setminus \{0\}$. Thus, $i((h(V) \cap S_\rho) \setminus N_\delta) \leq \dim X_2$. According to the subadditivity property of the $\mathbb{Z}_2$ index, we obtain

$$i(h(V) \cap S_\rho) \leq i((h(V) \cap S_\rho) \setminus N_\delta) + i(\bar{N}_\delta) \leq \dim X_2 + k. \tag{3.20}$$

On the other hand, since h and h^{-1} are bounded odd homeomorphism mappings, then $h(0) = 0$. Let $B_\rho = \{x \in X;\ ||x|| < \rho\}$. Thus, $h^{-1}(B_\rho) \cap V$ is a bounded open subset of V that contains the origin and is symmetric with respect to the origin, with its boundary denoted as Γ. Then, $\Gamma \subset V \cap h^{-1}(S_\rho)$. According to the Borsuk theorem and the monotonicity property of the $\mathbb{Z}_2$ index, we obtain

$$\dim V \leq i(\Gamma) \leq i(h^{-1}(S_\rho) \cap V) = i(h(V) \cap S_\rho).$$

Combined with (3.20), we have $k \geq \dim V - \dim X_2$. This is the end.

Theorem 3.21. *Let* $f \in C^1(X, \mathbb{R}^1)$ *be an even functional, and satisfy the following conditions*:

(1) $f(0) = 0$;
(2) *there exist* $\rho, \alpha > 0$*, and a closed linear subspace* X_1 *of* X*, such that the complement space* X_2 *of* X_1 *is* k*-dimensional, and* $f\Big|_{X_1 \cap S_\rho} \geq \alpha$;

(3) *there exist* m*-dimensional subspace* V *of* X*, and* $R > 0$*, such that* $m > k$*, and* $f(x) \leq 0$ *as* $x \in V \backslash B_R$;
(4) f *satisfies P.S. condition on* $f^{-1}((0, \infty))$.

Then f *possesses at least* $m - k$ *pairs of different critical points.*

Proof. Let H^* be the set composed of all bounded odd homeomorphism mappings, and $(\mathcal{A}, i_2^*)$ be the pseudo index of the $\mathbb{Z}_2$ index i with respect to H^*, where i_2^* is defined by (3.19). From Theorem 3.20, it is known that $i_2^*(V) \geq m - k$. Put

$$c_n^* = \inf_{i_2^*(A) \geq n} \sup_{x \in A} f(x), \quad n = 1, 2, \ldots.$$

Then, when $1 \leq n \leq m - k$, $\alpha \leq c_n^* < \infty$. From Theorem 3.17, it follows that f has at least $m - k$ pairs of different critical points. This is the end.

Corollary 3.3. *Suppose that the condition* (3) *in Theorem* 3.21 *is replaced by the following condition*:

(3′) X *is infinite dimensional, and for each finite dimensional subspace* V *of* X*, there exists* $R_V > 0$ *such that* $f(x) \leq 0$ *as* $x \in V$ *and* $||x|| \geq R_V$.

Then f *has an infinite pairs of different critical points.*

Corollary 3.4 (Symmetry Mountain Pass Theorem).
Suppose that the conditions (3) *and* (2) *in Theorem* 3.21 *are respectively replaced by* (3′) *in Corollary* 3.3 *and the following condition*:

(2′) *there exist* ρ *and* $\alpha > 0$*, such that* $f\Big|_{X \cap S_\rho} \geq \alpha$.

Then f *has an infinite pairs of different critical points.*

3.3.3 Category

To estimate the number of critical points, Ljusternik and Schnirelmann introduced the concept of category. This section only provides an introduction. For details, please refer to [6].

Definition 3.3.3. Let M be a topological space, a subset $F \subset M$ is said to be **contractible** in M, if the inject mapping $i : F \to M$

is homotopy with the constant mapping, that is, there exist $p \in M$ and $H \in C(F \times [0,1], M)$, such that $H(\cdot, 0) = i$ and $H(\cdot, 1) = p$.

Theorem 3.22. *The n-dimensional sphere S^n is not contractible in itself.*

Proof. Without loss of generalization, assume that S^n is the unit sphere. By contradiction, assume that there exists a homotopy mapping $\phi : S^n \times [0,1] \to S^n$, such that $\phi(u, 0) = u$, $\phi(u, 1) = u_0$, where u is any point on S^n and u_0 is a fixed point on S^n. Put

$$F(u) = \begin{cases} -u_0, & \text{when } u = 0, \\ -\phi(\frac{u}{||u||}, 1 - ||u||), & \text{when } 0 \neq u \in \bar{B}_{n+1}, \end{cases}$$

where $\bar{B}_{n+1}$ is the closed unit ball with center at 0 in $\mathbb{R}^{n+1}$. Since ϕ is uniformly continuous, $F : \bar{B}_{n+1} \to S^n$ is continuous. By the Brouwer fixed point theorem, there exists $u^* \in S^n$, such that $F(u^*) = u^*$. According to the definition of F, it is known that $F(u^*) = -u^*$, which is a contradiction. This is the end.

It is easy to see that S^n is contractible in $\mathbb{R}^{n+1}$. It can be proved in [6,27] that the unit sphere of an infinite dimensional Banach space is contractible in itself.

Definition 3.3.4. Let M be a topological space, A be a closed subset of M. Denote

$$\begin{aligned} \operatorname{cat}(A) = \inf\{ & m \in \mathbb{Z}_+ \cup \{\infty\}; \\ & \text{there exist } m \text{ contractible closed subsets } F_1, \ldots, F_m \\ & \text{such that } A \subset \textstyle\bigcup_{i=1}^m F_i\}, \end{aligned}$$

where $\mathbb{Z}_+$ is the set of nonnegative integers. $\operatorname{cat}(A)$ is called the **category** of A.

To emphasize that the definition of the category also depends on the topological space M itself, it is sometimes denoted by $\operatorname{cat}(A, M)$. It is easily known that if X is a Banach space, then $\operatorname{cat}(X, X) = 1$. Furthermore, it can be proven that if $S^n \subset \mathbb{R}^{n+1}$ is an n-dimensional sphere, then $\operatorname{cat}(S^n, S^n) = 2$.

The category possesses the following properties.

Theorem 3.23. (1) $\mathrm{cat}(A, M) = 0 \Longleftrightarrow A = \emptyset$.

(2) **(Monotonicity)** *If* $A \subset B \subset M$, *then* $\mathrm{cat}(A, M) \leq \mathrm{cat}(B, M)$.

(3) **(Subadditivity)** $\mathrm{cat}(A \cup B, M) \leq \mathrm{cat}(A, M) + \mathrm{cat}(B, M)$.

(4) **(Normality)** $\mathrm{cat}(\{p\}, M) = 1$, $\forall p \in M$.

(5) **(Deformation nondecreasing)** *Let h be homotopy with an identity mapping, that is, there exists a continuous mapping $\phi : M \times [0, 1] \to M$, such that $\phi(\cdot, 0) = id$, $\phi(\cdot, 1) = h$. Then*

$$\mathrm{cat}(A, M) \leq \mathrm{cat}(h(A), M).$$

(6) **(Continuity)** *If A is a compact subset of M, then $\mathrm{cat}(A, M) < \infty$, and there exists a closed neighborhood N of A, i.e., $A \subset \overset{\circ}{N} \subset N$, such that $\mathrm{cat}(A, M) = \mathrm{cat}(N, M)$.*

(7) *If $\mathrm{cat}(A, M) = m$, then A has at least m points.*

Proof. It is easy to see that the properties (1)–(4) follow from the definition directly.

(5) Assume $\mathrm{cat}(h(A), M) = m < \infty$. Then there exist closed contractible subsets $F_1, \ldots, F_m$, such that

$$h(A) \subset F_1 \cup F_2 \cup \cdots \cup F_m.$$

Let $G_i = h^{-1}(F_i)$, $i = 1, 2, \ldots, m$. Thus, G_i is closed, and $A \subset G_1 \cup G_2 \cup \cdots \cup G_m$. Since F_i is contractible, then there exists $\psi_i \in C(F_i \times [0, 1], M)$, such that $\psi_i(\cdot, 0) = id$ and $\psi_i(\cdot, 1) = p_i$ is constant. Setting $H_i(x, t) = t\psi_i(h(x), t) + (1-t)\phi(x, t)$, then G_i is contractible. So, $\mathrm{cat}(A, M) \leq m$.

(6) Firstly, we show a conclusion: every point in a Banach manifold M has a contractible neighborhood. Its proof can be found in [6, 27].

Next, we prove (6). Since A is compact, then there exist finite points $p_1, \ldots, p_k$, such that

$$A \subset U(p_1) \cup \cdots \cup U(p_k),$$

where $U(p_i)$ is a contractible neighborhood of the point p_i. Thus, $\mathrm{cat}(A, M) < \infty$.

Without loss of generalization, assume $\mathrm{cat}(A, M) = m$. Then there exist contractible closed subsets $F_1, \ldots, F_m$ such that $A \subset$

$\bigcup_{k=1}^{m} F_k$. For each F_k, there exists $\psi_k \in C(F_k \times [0,1], M)$ such that $\psi_k(\cdot, 0) = i$, $\psi_k(\cdot, 1) = p_k \in M$, where $i : F_i \to M$ is the inject mapping. By the extension theorem of the continuous mapping, there exist extensions $\tilde{\psi}_k : M \times [0,1] \to M$, such that $\tilde{\psi}_k\big|_{F_k \times [0,1]} = \psi_k\big|_{F_k \times [0,1]}$, $k = 1, \ldots, m$. Let V_k be a contractible closed neighborhood of p_k. Then $U_k := \tilde{\psi}_k^{-1}(V_k, 1)$ is closed and contractible. So, $U := \bigcup_{k=1}^{m} U_k$ is a closed neighborhood of A, that is, $m = \mathrm{cat}(A, M) \leq \mathrm{cat}(U, M) \leq m$.

(7) follows from the normality and subadditivity. This is the end.

The following theorem requires some knowledge of Banach manifolds. But, we do not want to elaborate on it. Please refer to [6, 7]. Let $\mathfrak{B}$ be the family of all closed subsets of M.

Theorem 3.24 (Ljusternik–Schnirelmann Theorem). *Let M be a C^{2-0} Finsler manifold, and $f \in C^1(X, \mathbb{R})$ satisfy P.S. condition. Denote*

$$c_k = \inf_{A \in \mathcal{F}_k} \sup_{x \in A} f(x), \quad k = 1, 2, \ldots,$$

where

$$\mathcal{F}_k = \{A \in \mathfrak{B};\ \mathrm{cat}(A, M) \geq k\}, \quad k = 1, 2, \ldots.$$

If

$$-\infty < c = c_{m+1} = c_{m+2} = \cdots = c_{m+k} < \infty,$$

then f has at least k different critical points in K_c.

Proof. According to the definition, $c_k \leq c_{k+1}, k = 1, 2, \ldots$. We only need to prove

$$\mathrm{cat}(K_c) \geq k.$$

By contradiction, assume that $\mathrm{cat}(K_c) = r \leq k - 1$. By the P.S. condition, K_c is compact. Due to the continuity of the category, there exists a closed neighborhood N of K_c, such that

$$\mathrm{cat}(K_c) = \mathrm{cat}(N).$$

From the definition of c, it follows that for any $\epsilon > 0$, there exists $F_\epsilon \in \mathcal{F}_{k+m}$ such that

$$\sup_{x \in F_\epsilon} f(x) < c + \epsilon.$$

According to the deformation theorem (the conclusion of the deformation theorem on Finsler manifolds is the same as the previous deformation theorem), there exists $\eta = \eta(1, \cdot)$ such that

$$\eta(F_\epsilon \setminus \mathring{N}) \subset \eta(f_{c+\epsilon} \setminus \mathring{N}) \subset f_{c-\epsilon},$$

which implies

$$\sup_{x \in \eta(F_\epsilon \setminus \mathring{N})} f(x) \leq c - \epsilon.$$

Because

$$\begin{aligned} k + m &\leq \mathrm{cat}(F_\epsilon) \leq \mathrm{cat}(F_\epsilon \setminus \mathring{N}) + \mathrm{cat}(N) \\ &\leq \mathrm{cat}(\eta(F_\epsilon \setminus \mathring{N})) + \mathrm{cat}(K_c) \\ &\leq \mathrm{cat}(\eta(F_\epsilon \setminus \mathring{N})) + k - 1, \end{aligned}$$

we obtain

$$\mathrm{cat}(\eta(F_\epsilon \setminus \mathring{N})) \geq m + 1.$$

So, we have

$$\sup_{x \in \eta(F_\epsilon \setminus \mathring{N})} f(x) \geq c_{m+1},$$

which is a contradiction. This is the end.

Corollary 3.5. *Let f satisfy the conditions in Theorem* 3.24, *and be bounded from below. Then f has at least* $\mathrm{cat}(M)$ *different critical points.*

Proof. By the assumptions, $c_k \leq c_{k+1}$. If $m = \mathrm{cat}(M)$ and $c_m < \infty$, it follows from the proof of the above theorem that there are at least m different critical points.

Suppose that there exists $m \leq \mathrm{cat}(M)$, such that $c_m = \infty$ and f only has a finite number of critical values (otherwise there are already an infinite number of critical points). Then, there must be a real number c, such that $K \cap f^{-1}([c, \infty]) = \emptyset$. By the deformation lemma, f_c is a deformation retraction of M. According to the deformation nondecreasing of the category, $m \leq \mathrm{cat}(f_c)$. This clearly implies

$$c_m = \inf_{\mathrm{cat}(A) \geq m} \sup_{x \in A} f(x) \leq c,$$

which is in contradiction with $c_m = \infty$. This is the end.

The remaining question is how to estimate $\mathrm{cat}(M)$. This is not an easy task. It involves some other topological invariants. For the interested readers, please refer to [6].

Exercises

1. Let $M = \{u \in C^1([1,2]);\ u(1) = 0, u(2) = 0\}$, and

$$I(u) = \int_1^2 \sqrt{1 + \dot{u}^2}\frac{dt}{t}.$$

   Find the function u that minimizes the functional I on M.
2. Find the minimum point of the functional

$$I(u) = \int_0^1 (t\dot{u} + \dot{u}^2)dt, \quad M = C_0^1(0,1).$$

3. Computer

$$\min\{I(u);\ u \in C^1[0,1], u(0) = 0, u(1) = 2, N(u) = L\},$$

   where

$$I(u) = \int_0^1 \dot{u}(t)^2 dt, \quad N(u) = \int_0^1 u(t)dt.$$

4. Let X be a real Banach space, and $f(x) = ||x||$. Show that $f(x)$ is weakly lower semi-continuous on X, that is, if $x_n \rightharpoonup x_0$ (weak), then $||x_0|| \leq \liminf_{n\to\infty} ||x_n||$.

5. Let $\Omega \subset \mathbb{R}^n$ be a bounded domain with smooth boundary, and $f \in C(\bar{\Omega})$. Denote the functional

$$I(u) = \int_\Omega \left[\frac{1}{4}|\nabla u|^4 - x^2|u|^2 - f(x)u\right] dx$$

on $W_0^{1,4}(\Omega)$. Show that the functional I is weakly lower semi-continuous and coercive, and has minimum point.
6. Let X be a Banach space, $f \in C^2(X, \mathbb{R})$, and $g(x) = \langle f'(x), x\rangle$. Suppose that $M = \{x \in X;\ g(x) = 0\}$ is a regular constraint, and $x_0 \in M$ is an extremum point of f restricted on M. If $\langle g'(x_0), x_0\rangle \neq 0$, show $f'(x_0) = 0$.
7. Let M be a closed convex subset of a Banach space X, and f be a C^1 functional on X. Suppose that x_0 is a local minimum point of f restricted on M. Show that x_0 solves the following variational inequality

$$\langle f'(x_0), x - x_0\rangle \geq 0, \qquad \forall x \in M.$$

8. Let $f \in C^1(\mathbb{R}^n, \mathbb{R})$ satisfy

$$||f'(x)|| + |f(x)| \to \infty, \quad ||x|| \to \infty.$$

Show that f satisfies the P.S. condition.
9. Let X be a Banach space, $f \in C^1(X, \mathbb{R})$ be bounded from below and satisfy the P.S. condition. Show that f is coercive, that is, $f(x) \to \infty$ as $||x|| \to \infty$.
10. Let X be a Banach space, and $f \in C^1(X, \mathbb{R})$ satisfy P.S. condition. Suppose that x_0 is a local minimum point of f, and there exists $x_1 \neq x_0$ such that $f(x_1) \leq f(x_0)$. Show that f has a critical point which is different from x_0.
11. Let X be a reflexive Banach space, and $f = f_1 + f_2$ be a coercive functional on X, where f_1 is weakly lower semi-continuous, and f_2 is weakly continuous. Show that f reaches its minimum value on X.
12. The symmetric bilinear functional F on the linear normed space X is called **nonnegatively definite**, if $F(h, h) \geq 0$ for any $h \in X$. Suppose that f is a twice continuously differentiable functional on X, and f reaches a local minimum at $x_0 \in X$. Show that $f''(x_0)$ is nonnegatively definite.

13. Let X be a Banach space, $f \in C^1(X, \mathbb{R})$ and satisfy the P.S. condition on $f^{-1}((a,b))$. Assume $c \in (a,b)$ such that $K_c = \emptyset$. Show that there exists $\bar{\epsilon} > 0$ such that

$$K \cap f^{-1}([c - \bar{\epsilon}, c + \bar{\epsilon}]) = \emptyset.$$

14. Replace $\Phi(x)$ in the proof of deformation lemma with $\Phi(x) = -\chi(x)\tilde{\chi}(x)\frac{V(x)}{||f'(x)||^2}$. Provide a proof of the deformation lemma.
15. Prove that when f is an even functional, the mapping in the deformation lemma can be taken as an odd mapping.
16. Let $H = W^{1,2}(\mathbb{R})$, $J : H \to \mathbb{R}$, $J(u) = \int_{-\infty}^{\infty} \left[\frac{1}{2}(\dot{u}^2 + u^2) - \frac{1}{4}u^4\right] dx$. Prove that the P.S. condition does not hold.
Hint: Consider $c = \frac{4}{3}$.
17. Let $f(x,y) = x^2 + (x+1)^3 y^2$. Show that
    (1) $f(0,0)$ is a strictly local minimum;
    (2) there exist $\rho > 0$ and $\alpha > 0$, such that $f(x,y) \geq \alpha$ as $x^2 + y^2 = \rho^2$;
    (3) there exists a point $(x_0, y_0) \in \mathbb{R}^2$ with $x_0^2 + y_0^2 > \rho^2$, such that $f(x_0, y_0) \leq 0$;
    (4) the P.S. condition is not satisfied;
    (5) the conclusion of the mountain pass lemma is also not valid, that is, the c defined in the mountain pass lemma is not a critical value.

Bibliography

[1] R. Adams and J. Fournier, *Sobolev Spaces*, Second Edition, Elsevier (Singapore) Pte Ltd., 2009.
[2] J. Andres and L. Górniewicz, *Topological Fixed Point Principles for Boundary Value Problems*, Kluwer Academic Publishers, Dordrecht/Boston/London, 2003.
[3] K. C. Chang, *Methods in Nonlinear Analysis*, Springer-Verlag, 2005.
[4] K. C. Chang and Y. Q. Lin, *Lectures on Functional Analysis I (Chinese)*, Peking University Press, 1987.
[5] K. C. Chang and M. Z. Guo, *Lectures on Functional Analysis II (Chinese)*, Peking University Press, 1990.
[6] K. C. Chang, *Critical Theory and Its Applications (Chinese)*, Shanghai Science and Technology Press, 1986.
[7] K. C. Chang, *Lectures on Variational Methods (Chinese)*, Higher Education Press, 2011.
[8] S. N. Chow and J. K. Hale, *Methods of Bifurcation Theory*, Springer-Verlag, 1982.
[9] P. G. Ciarlet, *Linear and Nonlinear Functional Analysis with Applications*, SIAM, 2013.
[10] K. Deimling, *Nonlinear Functional Analysis*, Springer-Verlag, 1985.
[11] R. Gaines and J. Mawhin, *Coincidence Degree and Nonlinear Differential Equations*, LNM Vol 568, Springer-Verlag, 1977.
[12] A. Granas and J. Dugundji, *Fixed Point Theory*, Springer-Verlag, New York, 2003.
[13] D. J. Guo, *Nonlinear Functional Analysis (Chinese)*, Shandong Science and Technology Press, 1985.
[14] B. H. He and G. F. Liao, *Fundamental Topology (Chinese)*, Higher Education Press, 1991.

[15] G. Iooss and D. Joseph, *Elementary Stability and Bifurcation Theory*, Springer-Verlag, 1980.
[16] W. Krawcewicz and J. Wu, *Theory of Degrees with Applications to Bifurcations and Differential Equations*, John Wiley & Sons, Inc. New York, 1997.
[17] Y. M. Long, *Index Theory of Hamiltonian Systems and Its Applications (Chinese)*, Science Press, 1993.
[18] J. Mawhin and M. Willem, *Critical Point Theory and Hamiltonian Systems*, Springer-Verlag, 1989.
[19] L. Nirenberg, *Topics in Nonlinear Functional Analysis*, Courant Lecture Notes 6, AMS, 2001.
[20] J. Smoller, *Shock Waves and Reaction-Diffusion Equations*, Springer-Verlag, 1994.
[21] M. Struwe, *Variational Methods, Applications to Nonlinear Partial Differential Equations and Hamiltonian Systems*, Second Edition, Springer-Verlag, 1996.
[22] J. X. Sun, *Nonlinear Functional Analysis and Its Applications (Chinese)*, Science Press, 2008.
[23] D. X. Xia, S. Z. Yan, W. C. Shu and Y. S. Tong, *Functional Analysis Second Tutorial (Chinese)*, Second Edition, Higher Education Press, 2008.
[24] D. C. Yang and W. Yuan, *Selected Lectures on Functional Analysis (Chinese)*, Beijing Normal University Press, 2016.
[25] E. Zeidler, *Applied Functional Analysis*, Springer-Verlag, 1995.
[26] E. Zeidler, *Nonlinear Functional Analysis and Its Applications I, II, III, IV*, Spring-Verlag, 1986.
[27] F. B. Zhang, *Fundamentals of Modern Analysis and Its Applications (Chinese)*, Science Press, 2014.
[28] C. K. Zhong, X. L. Fan and W. Y. Chen, *Nonlinear Functional Analysis (Chinese)*, Lanzhou University Press, 2004.

Index

www.ingramcontent.com/pod-product-compliance
Ingram Content Group UK Ltd.
Pitfield, Milton Keynes, MK11 3LW, UK
UKHW062100120926
8588IPUK00007B/64

9 789819 825813